Camellias

Chang Hung Ta

Bruce Bartholomew

TIMBER PRESS
Portland, Oregon
1984
in cooperation with
THE AMERICAN CAMELLIA SOCIETY
Ft. Valley, Georgia

© Timber Press 1984

All rights reserved.
Printed in the United States of America.
ISBN 0-917304-81-0

TIMBER PRESS
P.O. Box 1631
Beaverton, Oregon 97075

Contents

Preface	1
Chapter 1 INTRODUCTION	13
Taxonomy	13
Phylogeny	16
Geographic Distribution	18
Economic Use	24
Chapter 2 CAMELLIA	27
Chapter 3 SUBGENUS *PROTOCAMELLIA*	31
Section I *Archecamellia*	31
Section II *Stereocarpus*	32
Section III *Piquetia*	36
Chapter 4 SUBGENUS *CAMELLIA*	37
Section IV *Oleifera*	37
Section V *Furfuracea*	40
Section VI *Paracamellia*	48
Section VII *Pseudocamellia*	63
Section VIII *Tuberculata*	67
Section IX *Luteoflora*	73
Section X *Camellia*	74
Chapter 5 SUBGENUS *THEA*	111
Section XI *Corallina*	111
Section XII *Brachyandra*	116
Section XIII *Longipedicellata*	125
Section XIV *Chrysantha*	128
Section XV *Calpandria*	136
Section XVI *Thea*	137
Section XVII *Longissima*	153
Section XVIII *Glaberrima*	154
Chapter 6 SUBGENUS *METACAMELLIA*	157
Section XIX *Theopsis*	157
Section XX *Eriandria*	199
Index	209

ILLUSTRATIONS

Figures
1. Phylogenetic tree of *Camellia*   17
2. *Camellia* species and isotherms within China   20
3. Hardiness zones of North America   21
4. Hardiness zones of Europe   22

Line drawings
1. *C. granthamiana*   30
2. *C. albogigas*   33
3. *C. yunnanensis*   35
4. *C. vietnamensis*   39
5. *C. integerrima*   42
6. *C. oblata, C. polypetala*   44
7. *C. crapnelliana*   46
8. *C. parafurfuracea*   47
9. *C. fluviatilis, C. kissii*   52
10. *C. brevistyla*   53
11. *C. obtusifolia*   55
12. *C. shensiensis*   58
13. *C. tenii, C. puniceiflora, C. microphylla*   59
14. *C. phaeoclada*   61
15. *C. weiningensis*   62
16. *C. chungkingensis*   65
17. *C. henryana*   68
18. *C. tuberculata, C. szechuanensis*   69
19. *C. parvimuricata*   72
20. *C. omeiensis*   78
21. *C. polyodonta*   79
22. *C. lapidea, C. longicaudata*   81
23. *C. mairei*   82
24. *C. kweichowensis*   84
25. *C. albovillosa*   85
26. *C. tunganica*   87
27. *C. phellocapsa*   89
28. *C. multiperulata, C. lungshenensis*   91
29. *C. reticulata*   93
30. *C. hongkongensis, C. japonica*   96
31. *C. cryptoneura*   98
32. *C. compressa*   100
33. *C. setiperulata*   101
34. *C. saluenensis, C. hiemalis*   103
35. *C. chekiangoleosa*   105
36. *C. lucidissima, C. subintegra*   106
37. *C. lienshanensis*   115
38. *C. pachyandra, C. pilosperma*   119
39. *C. brachyandra*   121
40. *C. parviflora*   124
41. *C. glaberrima, C. longipedicellata*   126
42. *C. indochinensis*   127
43. *C. chrysantha*   131
44. *C. euphlebia, C. tunghinensis*   133
45. *C. pingguoensis*   135
46. *C. kwangsiensis*   140
47. *C. gymnogyna, C. yungkiangensis*   145
48. *C. angustifolia*   148
49. *C. parvisepala*   152
50. *C. kwangtungensis*   156
51. *C. macrosepala*   161
52. *C. longicuspis*   164
53. *C. longicalyx*   165
54. *C. forrestii*   167
55. *C. acutissima*   170
56. *C. subacutissima*   171
57. *C. handelii*   173
58. *C. costei*   174
59. *C. tsaii, C. trichoclada*   176
60. *C. tsaii* var. *synaptica*   177
61. *C. rosthorniana*   179
62. *C. parvilimba*   180
63. *C. elongata*   182
64. *C. cratera, C. longicarpa*   183
65. *C. parvilapidea*   185
66. *C. fraterna*   186
67. *C. dubia*   187
68. *C. campanisepala*   190
69. *C. lancilimba*   191
70. *C. viridicalyx, C. parvicaudata*   194
71. *C. trigonocarpa, C. villicarpa, C. tsofuii*   197
72. *C. trichandra*   198
73. *C. cordifolia, C. caudata*   203
74. *C. edentata*   206

Preface

The present work is basically a translation from Chinese to English of the 1981 monograph of the genus *Camellia* by H. T. Chang (Chang 1981). However, in cooperation with Professor Chang, the translation has been updated to include publication of new taxa through the end of 1982. There have been some other additions and changes to the original work including correction of typographic errors, changes necessary to make the English translation of the Chinese agree with the Latin descriptions, lectotypification of a few species, publication of species that were not validly published in the original work, addition of information on the location of holotypes of some of the previously published species, and citation of additional synonyms. Also, Professor Chang has been able to supply many more line drawings than were in the original work. All of the additions and changes were done in conjunction with Professor Chang.

Only the Chinese has been translated into English. The Latin descriptions and diagnoses of the original work have not been changed except as mentioned above. The inclusion of the Latin descriptions and diagnoses was deemed important both to make the present work complete and because the original work, where the Latin is required by the rules of botanical nomenclature, is available only to a limited extent outside of China.

Although almost all the species in the genus *Camellia* are treated in Professor Chang's monograph, the descriptions are not complete for some of the older and well known species such as *C. japonica, C. oleifera, C. pitardii, C. reticulata* and *C. sinensis*. For more complete descriptions of these species the reader is referred to the Sealy's monograph (Sealy 1958).

As Sealy's work is the most recent treatment of the genus, it is of interest to compare the infrageneric classification of the two works as was reported by Bartholomew (1981). In Sealy's monograph there are 12 sections containing 82 species as well as a group of 24 species that Sealy called *"Dubiae"*. The species in this group were too inadequately known for Sealy to place them in any of the 12 sections. Chang has retained all of the sections in Sealy's work except for section *Heterogena*. However, many species treated by Sealy have been moved to different sections.

Chang emmends section *Archecamellia* to exclude the nomenclatural type, *C. petelotii*, and *C. granthamiana* is designated as the new type of this section. Of the species recognized by Sealy in section *Archecamellia* only *C. pleurocarpa* remains. The other species are moved to sections *Stereocarpus (C. krempfii), Corallina (C. tonkinensis), Longipedicellata (C. amplexicaulis* and *C. petelotii)* and *Chrysantha (C. flava* and *C. euphlebia)*.

Both *C. sasanqua* and *C. oleifera* are moved from section *Paracamellia* to the new section *Oleifera* while *C. oleifera* var. *confusa* is raised to *C. confusa* and remains in section *Paracamellia*.

C. tuberculata is moved from section *Pseudocamellia* to a new section *Tuberculata*. The four species *C. amplexifolia, C. nematodea, C. gilbertii* and *C. parviflora* are moved from section *Corallina* to a new section *Brachyandra*. *C. gracilipes* is moved from section *Thea* to a new section *Longissima*.

Six species that Sealy treats in section *Theopsis* are moved to different sections. These changes include the move of *C. indochinensis* to section *Longipedicellata, C. maliflora* to section *Paracamellia*, and *C. assimiloides, C. lawii, C. punctata* and *C. villicarpa* to section *Eriandria*. It should be pointed out that both Sealy (1958) and Chang (1981) use section

Camelliopsis, but by the rules of botanical nomenclature the ending *opsis* cannot be added as the ending of a genus to form the name of a section within the same genus. This problem was pointed out by Bartholomew (1981) and is the reason for the use of the section *Eriandria* in the translation rather than section *Camelliopsis*.

In his 1958 monograph Sealy formed section *Heterogena*, but he points out that, "As its name implies, this section is a collection of diverse species . . . , and since they do not fit easily into any other groups of species, it is convenient to put them together in a separate section." This section is completely eliminate by Chang and the species are distributed among other sections which include *Archecamellia* (*C. granthamiana*), *Stereocarpus* (*C. yunnanensis*), *Furfuracea* (*C. crapnelliana, C. furfuracea* and *C. gaudichaudii*), *Paracamellia* (*C. tennii*), *Pseudocamellia* (*C. henryana*) and *Corallina* (*C. wardii, C. fleuryi* and *C. paucipunctata*).

Of the twenty-four species included by Sealy in his group *"Dubiae"* Chang distributes ten into sections *Furfuracea* (*C. latipetiolata*), *Paracamellia* (*C. lutescens, C. miyagii* and *C. microphylla*), *Camellia* (*C. semiserrata, C. hiemalis* and *C. uraku*), *Corallina* (*C. nitidissima*) and *Brachyandra* (*C. pachyandra* and *C. nervosa*). The remaining 14 species are not treated by Chang.

The line drawings in this volume are by the following artists: J. Deng, 8, 14, 16, 39, 46, 48, 50, 53, 55, 58, 68, 69; X. R. Huang, 31, 56, 72, 74; K. K. Sie, 3, 4, 7, 9, 15, 20, 21, 22, 30, 32, 35, 42, 43, 45, 52, 54, 57, 60, 61, 62, 63, 66, 67, 73; S. C. Tsai, 1, 2, 6, 10, 12, 13, 17, 18, 23, 25, 26, 28, 29, 33, 34, 36, 38, 41, 44, 47, 59, 64, 70, 71; F. Yu, 5, 24, 37, 51; H. Yu, 11, 40, 49, 65.

The acronyms for herbaria cited in the publication follows those listed in *Acta Phytotax. Sin.* (Anonymous 1982a). These are as follows:

A (Arnold Arboretum of Harvard University)
E (Royal Botanic Garden, Edinburgh)
GXFI (Guangxi Institute of Forestry)
GXMI (Guangxi Institute of Traditional Medical & Pharmaceutical Sciences)
HLG (Lushan Botanical Garden)
HNTC (Hunan Teacher's College)
HZBG (Hangzhou Botanic Garden)
JSBI (Jiangsu Institute of Botany)
K (Royal Botanical Gardens, Kew)
KUN (Kunming Institute of Botany, Academia Sinica)
NA (National Arboretum, Washington, D.C.)
NJTFC (Nanjing Technological College of Forest Products)
NJU (Nanjing University)
PE (Institute of Botany, Academia Sinica)
SCBI (South China Institute of Botany, Academia Sinica)
SYS (Sun Yatsen University)
SZ (Sichuan University)
YNTBI (Yunnan Institute of Tropical Botany, Academia Sinica)

THE EARLY INTRODUCTION OF *CAMELLIA* TO THE WEST

Although *Camellia* has a long history of cultivation in Asia, its cultivation in the West dates back only about 250 years. As a result of this relatively recent introduction the early history of *Camellia* in the West is fairly well known (Booth 1830, Bretschneider 1880, 1898; Sealy 1937, Hume 1955, Sealy 1958, Brown 1978). During the period from the mid 18th century to the end of the 19th century, nine *Camellia* species were brought into cultivation in the West. A tenth, *C. hongkongensis*, was briefly grown at Kew but did not become established during the 19th century (Sealy 1958). Before 1900 additional species were described including *C. assimilis, C. caudata, C. dormoyana, C. edithiae, C. fraterna, C. grijsii, C. lanceolata, C. lutescens, C. piquetiana* and *C. salicifolia*. These species were either brought into cultivation in the West after 1900 or in some cases are still not yet cultivated.

The earliest species to be cultivated in the West was *Camellia japonica*. Although the earliest illustration of this species was by James Petiver (1702–1709) based on herbarium specimens collected by James Cunningham in the late 17th century, the first live introduction of this species to the West was a plant grown by James Lord Petre at least as early as 1739 (Aiton 1789, 1810-13, Booth 1830). Lord Petre was the 8th Baron Petre of Writtle and a fellow of the Royal Society as well as an avid horticulturist with a garden at Thorndon Hall, Essex. Unfortunately he died of smallpox at the age of 29 in 1742 (Henrey 1975). The first color illustration of a *Camellia* in the West (Edwards 1747) is that of a *C. japonica* but using the name Chinese rose. *C. japonica* did not start becoming a major cultivated species until later in the 18th century, and probably the two most important early introductions were 'Alba Plena' and 'Variegata' which were brought to England in 1792 by Captain Connor of the East India Merchantman Carnatic for John Slater (Andrews 1797, Booth 1830, Don 1831). Over the next few years more camellias were imported, and by 1812 eleven varieties are listed as being grown at Kew (Aiton 1810–1813). During the early to mid 19th century many more cultivars were grown in England and Western Europe as can be seen from the numerous color plates published in horticultural magazines such as *The Botanical cabinet, consisting of coloured delineations of plants from all countries...* by Conrad Loddiges & Sons, *The Botanical register*, *The Botanists' repository, for new and rare plants...* by Henry C. Andrews, *Curtis's botanical magazine* and *Flores des serres et des jardins de l'Europe* as well as lavishly illustrated books on *Camellia* such as those by Curtis (1819), Booth & Chandler (1831), and Berlèse (1841–43). Although many new cultivars of *C. japonica* were selected from seedlings grown in England and Western Europe during this period, most of the introductions from Asia were from Chinese gardens. There are two reason that the early introductions from Asia came from China rather than Japan. Japan was largely closed to foreign contact between the early 17th century to the mid 19th century, and *C. japonica* is a more important garden plant in China than in Japan where *C. sasanqua* is still more prevalent. Many of these early cultivars probably still exist in China. For example the late 18th century introduction, 'Alba Plena', is still commonly grown in China, and in Yunnan Province is often used as an understock for cultivars of *C. reticulata*.

Camellia sinensis was the second species to be brought into cultivation in the West. This species was apparently cultivated in England as early as 1740 by Captain Goff of the East India Company, but this introduction died (Sealy 1958). The earlier literature claims that the first successful introduction was by John Ellis in 1768 (Aiton 1789, Booth 1830) who was a merchant and naturalist as well a member of the Royal Society and correspondent with Linnaeus. However, the first successful introduction appears to have been done for Linnaeus by Captain Carolus Gustavus Ekeberg of the Swedish East India Company in 1763 (Linnaeus 1766–1767, Bretschneider 1898, Sealy 1958). Seeds were planted shortly after leaving China so that seedlings were already growing by the time Captain Ekeberg arrived back in Europe. As the source of tea, *C. sinensis* is by far the most important economic plant within the genus, but although it has been grown as an ornamental plant in the West for over two hundred years, it has never been an important ornamental plant.

C. oleifera was described and illustrated by Dr. Clarke Abel (1818) who was appointed the naturalist attached to Lord Amherst's Embassy to the Chinese Court, 1816–17. It is possible that this species is the same one named *C. drupifera* by Ioannis de Loureiro in 1790 in his *Flora Cochinchinensis*. The reason the older name is not used is because of uncertainty over the identity of Loureiro's species. Unfortunately Loureiro's descriptions are rather cryptic, and the specimen on which he based his name is not extant. *C. oleifera* was first introduced to the West by William Kerr, who was appointed botanical collector for Kew in Canton in 1803 (Bretschneider 1898). *C. oleifera* was sent back to England in 1811 as *C. sasanqua* and grown at Kew as 'Lady Bank's Camellia' (Aiton 1810–1813). During the 19th century there was frequently confusion in the horticultural literature between *C. oleifera* and *C. sasanqua*. The true *C. sasanqua* was not grown in England until much later.

Another species first confused with *C. sasanqua* is *C. maliflora*, first described as *C. sasanqua* 'Palmer's Double' by John Sims (1819) based on a plant brought from China by Captain Richard Rawes for Thomas Carey Palmer. This plant was subsequently recognized by Lindley as a distinct species (Lindley 1827). *C. maliflora* is a double flowered *Camellia* that has never been found in the wild and is most probably of hybrid origin.

The next species to be introduced to the West was *C. reticulata*. A semi-double form of this species was introduced to England in 1820, again for Palmer by Captain Rawes. The species was named in 1827 on the basis of what was probably the same cultivar introduced to England in 1824 by J. D. Parks (Lindley 1827) and given the English name 'Captain Rawes's Camellia' in honor of the Captain who brought it to England. We now know that this species is native to Yunnan Province, and that the plants introduced to England were no doubt sent down from Yunnan to the flower markets of Guangzhou or Macau. *C. reticulata* blooms around the time of the Chinese New Year, and plants of this species are still sent down to the flower markets of Hong Kong from Yunnan for the lunar New Year festival. The cultivar 'Captain Rawes' has not been identified with any of the *C. reticulata* cultivars still growing in Yunnan Province, although the cultivar must have been grown in China for more than a hundred years before its introduction to England if, as has been argued by Yoshiaki Andoh, this same cultivar was introduced from China to Japan in the late 17th century (Andoh 1972). A second cultivar of *C. reticulata* was brought back to England by Robert Fortune, probably during his second expedition to China (1848–1851) and grown at the nursery of Standish and Noble. Fortune's introduction was given the name 'Flore Pleno' (Lindley 1857). The name 'Robert Fortune' was later used for the cultivar, and on the basis of material introduced to the United States from Yunnan Province by Ralph Peer and W. E. Lammerts in the late 1940s, it has been determined to be the same as that known in China as 'Songzilin' or "pine-cone-scale". For almost a hundred years *C. reticulata* was known in the West only by these two cultivars. It was not until the early 1930s that some of the plants grown from seeds collected between 1913 and 1925 by the Scottish plant explorer George Forest were recognized to be wild forms of *C. reticulata* (Sealy 1935). Wild growing specimens collected by George Forest all came from a very limited area in the vicinity of Tengchong (formerly called Tengyueh) in extreme western Yunnan Province near the Burmese border. In addition to these wild plants, this species has long been cultivated in Yunnan Province, particularly in and around the cities of Kunming and Dali (Hu 1938, Yü 1950, Yü and Feng 1958, Yü 1964, Yü and Bartholomew 1980, Feng *et al.* 1981, Bartholomew 1982), where many cultivars have been selected and named.

C. euryoides has an unusual origin. It was described by Lindley in 1826 from a plant brought back to England by John Potts, who collected in China and India in 1821 and 1822 for the Horticultural Society of London (later to be the Royal Horticultural Society) (Bretschneider 1898). *C. euryoides* was the rootstock onto which a *C. japonica* cultivar had been grafted by Chinese gardeners. The *C. japonica* cultivar died while the rootstock grew and flowered for the first time in March 1826 in the Cheswick Garden (Lindley 1926). That this species was not an unusual understock for Chinese gardeners to use can be seen by *C. euryoides* appearing again from the understock of a plant brought back to England by Parks in 1824 (Lindley 1826). This species has subsequently been found growing in the wild in southeastern China including Fujian, Guangdong and Jiangxi provinces (Chang 1981).

C. kissii was described in 1820 by Wallich and was introduced into cultivation by Samuel Brooks in 1823 (Booth 1830, Don 1831–1838). This species does not seem to have attracted much attention from the horticultural community at the time. It is probable that plants of this species now grown in the West are from more recent introductions.

C. rosaeflora was of completely unknown provenance when it was described by Hooker in 1858. The description was based on a plant of unknown origin that had long been grown at Kew as *C. euryoides* but was recognized by Hooker as a previously undescribed species. This species was subsequently lost in cultivation in England but was rediscovered as a cultivated plant in Ceylon in 1935 and reintroduced to England in 1956 (Sealy 1958). *C. rosaeflora* is now known to be a wild species, and Chang (1981) gives the distribution of *C. rosaeflora* in China as including Jiangsu, Zhejiang, Sichuan and Hubei provinces.

The first appearance of *C. sasanqua* as a cultivated plant in the West was in a color illustration in 1869 published in France (Lemaire 1869). No mention is made in the article as to the origin of the illustration, but there can be no doubt as to the identity of the species. Although Brown (1978) believes that the plant grown in Italy in the royal gardens at Caserta near Naples as early as 1760 and mentioned by Berlèse (1837) was a *C. sasanqua*, this plant was probably *C. japonica* as stated by Linden and Rodigas (1886). *C. sasanqua* was known by Kaempfer in the late 17th century as it was published in his *Amoenitatum Exoticarum Politico-Physico-Medicarum* (Kaempfer 1712) based on observations during his stay in Japan between 1690 and 1692. Kaempfer used the name *Sasanqua* but did not illustrate the plant. Linnaeus does not mention this species in his *Species Plantarum* (Linnaeus 1753), but used Kaempfer's descriptions as the basis for his publication of *Thea sinensis* and *Camellia japonica*. Had Kaempfer illustrated the plant he called *Sasanqua* as he did with the other two *Camellia* species, it is probable that this third species would have been included in Linnaeus' *Species Plantarum*, but the species was not formally recognized until Thunberg, a student of Linnaeus, published the binomial *C. sasanqua* in his *Flora Japonica* (Thunberg 1784). *C. sasanqua* was first grown in England in 1879 from plants collected in Japan by Charles Maries who was a plant collector for the nursery of James Veitch & Sons, and Sealy (1937, 1958) believes that all the plants grown in England as *C. sasanqua* before this date were misidentified, mostly being forms of *C. oleifera* or the species *C. maliflora*. It is not surprising that *C. sasanqua* was introduced relatively late to the West, because Japan was largely closed to foreigners before the mid 19th century.

Since the beginning of the 20th century additional species have been introduced to the West, but there are still many wild growing species and no doubt many selected forms of garden grown plants that would add greatly to ornamental horticulture in the West. There are also a great many *Camellia* cultivars that have been selected in the West that would add greatly to the gardens in Asia. Hopefully the exchange of plants between the East and West will enhance the quality of life for the people of both areas.

THE CULTIVATED SPECIES

There are several *Camellia* species known only from cultivated plants. As treated in the present work these species include *C. hiemalis, C. maliflora* and *C. uraku*. Although not treated in this work *C. wabiske* (Mak.) Kitamura and *C. vernalis* (Mak.) Mak. should also be mentioned.

Chang places *C. hiemalis* in Section *Camellia* and teats it as a species. However, there is evidence that both *C. hiemalis* and *C. vernalis* are hybrids involving *C. japonica* and *C. sasanqua* (Parks et al. 1967, Uemoto et al. 1980, Parks et al. 1981). In addition Kondo (1976) presents evidence that *C. hiemalis* did not come from China in 1925, as was indicated by Nakai (1940), but probably originated in Japan. If this is the case, the cultivated plants of *C. hiemalis* in China probably came from Japan rather than the other way around.

Parks *et al.* (1981) also present phytochemical evidence that many of the cultivated plants of *C. sasanqua* are derived from hybrids between *C. sasanqua* and *C. japonica* that have introgressed back to *C. sasanqua*. As compared with wild growing plants, the morphological and phenological evidence of genetic contamination of *C. sasanqua* by *C. japonica* includes, "a tendency to bloom later in the autumn or early in the winter, some fusion of petal and filament bases, pink and red shades of floral pigmentation and relatively coarser branch and foliage texture" (Parks *et al.* 1981).

From the standpoint of nomenclature, *C. hiemalis, C. vernalis* and those cultivars of *C. sasanqua* that show evidence of hybridization with *C. japonica* should all be treated as synonymous under a single hybrid specific epithet. Unless either one of the earlier synonyms of *C. sasanqua* (i.e. *Sasanqua vulgaris* or *S. odorata*) or the holotype of *C. sasanqua* can be shown to be hybrids with *C. japonica*, the name that should be applied to these hybrids is *C.* X *vernalis.*

C. maliflora is another species that is only known in cultivation. Chang places this species in section *Paracamellia* but considers the species to be a hybrid. It would seem better to use the name *C.* X *maliflora* and not place the hybrid epithet in an infrageneric classification.

In 1910 the Japanese botanist Makino described four varieties of cultivated *Thea* (Makino 1910) which he later transferred to *Camellia* (Makino 1918). These plants included *C. reticulata* var. *rosea* (Mak.) Mak., *C. reticulata* var. *albo-rosea* (Mak.) Mak., *C. reticulata* var. *wabiske* (Mak.) Mak. and *C. reticulata* var. *campanulata* (Mak) Mak. These varieties have also been treated as forms of *C. wabiske* (Mak.) Kitamura (Kitamura 1970). In the case of *C. reticulata* var. *rosea,* the variety was first treated at the specific level as *C. uraku* Kitamura (Kitamura 1952) so when it was reduced to a form of *C. wabiske* the name *C. wabiske* f. *uraku* (Kitamura) Kitamura was used. Chang treats *C. uraku* as a species in section *Camellia* but does not treat the other three taxa. The correct taxonomic status of these camellias is still uncertain. They form a group of cultivated camellias of which the original stock is thought to have been brought to Japan from China 350 to 400 years ago (Kitamura 1952, Kondo 1976). Whether all of these taxa are interspecific hybrids or whether any of them should be considered as distinct species must wait additional phytochemical and cytogenetic work.

HORTICULTURAL POTENTIAL

The genus *Camellia* has long been important in horticulture, yet very few species have been involved. By far the most important is *C. japonica* with several thousand named cultivars. The other horticulturally important species include *C. sasanqua* and *C. reticulata*. Although there are cultivars of other species these three species and their hybrids form the vast majority of cultivars. However, additional species are in cultivation and in the 1981 edition of *Camellia Nomenclature* (W. E. Woodroof, ed. 1981) 24 species are listed as being in cultivation in the United States. Although some of these specific epithets are probably interspecific hybrids, this list includes the following: *C. caudata, C. connata, C. cuspidata, C. fraterna, C. granthamiana, C. hiemalis, C. hongkongensis, C. irrawadiensis, C. japonica, C. maliflora, C. oleifera, C. pitardii, C. reticulata, C. rosaeflora, C. rusticana* (usually considered to be a subspecies of *C. japonica*), *C. salicifolia, C. saluenensis, C. sasanqua, C. sinensis, C. taliensis, transnokoensis, C. tsaii, C. vernalis* and *C. wabiske.*

To this list can be added the following 18 species that have brought into cultivation comparatively recently (Anonymous 1981, 1982b): *C. assimilis, C. brevistyla, C. chekiangoleosa, C. chrysantha, C. crapnelliana, C. forrestii, C. furfuracea, C. grijsii, C. kissii, C. lutchuensis, C. miyagii, C. nokoensis, C. octopetala* (now reduced to a synonym of *C. crapnelliana*), *C. parviflora, C. transarisanensis, C. vietnamensis, C. yuhsienensis* (now reduced to a synonym of *C. grijsii*) and *C. yunnanensis.* Although there are probably a few additional unreported species in cultivation in the West and certainly additional species in cultivation in China and Japan not yet grown in the West, there are now in excess of 40 *Camellia* species known to be in cultivation. However, this total is still only about twenty percent of the species treated in the present work.

It is probable that in the foreseeable future it will be impossible to bring all of the approximately 200 *Camellia* species into cultivation. Some of the species are exceedingly rare even in the wild or known only from a single collection. In addition, although some botanical gardens would no doubt be interested in growing all the known species, many species are certainly of limited horticultural value. In particular, probably a comparatively few of the 57 species in subgenus *Metacamellia* will ever be of major importance as horticultural plants.

If past history is to be a guide, the most important groups for the future will come from subgenus *Camellia* and particularly section *Camellia*. As treated by Chang, within this section are species already in cultivation such as *C. chekiangoleosa, C. hiemalis, C. hongkongensis, C. japonica, C. pitardii, C. saluenensis* and *C. uraku*. However, this still leaves 25 species not yet in cultivation in the West. All of these species would be of value to bring into cultivation both for their own merit as well as for breeding into horticulturally important species such as *C. japonica* and *C. reticulata*.

After section *Camellia* the other two most important sections, both in terms of present and future use are the species in the closely related sections *Oleifera* and *Paracamellia*. In addition to its importance as an oil source, *C. oleifera* has already been shown to be important for breeding cold hardiness into cultivated camellias (Ackerman 1978). In this regard *C. shensiensis* is also of potential importance due to its northerly distribution.

Sections *Oleifera* and *Paracamellia* are also important as potential breeding sources for flower fragrance. In addition to *C. sasanqua, C. grijsii* will no doubt be an important source for breeding fragrance into other *Camellia* species. *C. grijsii* is quite variable in terms of flower fragrance, and the plants that are the most fragrant are those formerly known as *C. yuhsienensis* (Zhang Aoluo, per. com.).

Species of section *Furfuracea* will undoubtedly find an important place as cultivated plants, but not for the aesthetic value placed on the flowers as in section *Camellia*. The species in section *Furfuracea* tend to have large coriaceous leaves and will no doubt be introduced because they will make handsome shrubs and small trees under cultivation. Their qualities of form and color suiting them for cultivation has already been commented upon by Parks (1982).

Although the species in subgenus *Protocamellia* occur within the tropics and as such will be difficult to grow in temperate areas, all of the eight species in this subgenus appear to be admirably suited as horticultural plants in subtropical and tropical growing areas. At present only *C. granthamiana* and *C. yunnanensis* are in cultivation outside of Asia. The other six species should be introduced as they would certainly add greatly to gardens in subtropical and warm temperate parts of the world.

Thea is of course by far the most important subgenus economically, but in cultivation is almost exclusively restricted to *C. sinensis*. The flowers in the subgenus are characteristically small to medium sized, and as such are not as showy as those of many species in subgenera *Camellia* or *Protocamellia*. The one section in this subgenus that has received the most attention in recent years is *Chrysantha* and particularly *C. chrysantha*.

It has long been the dream and goal of many *Camellia* breeders to develop yellow *Camellia* flowers. The problem now is to breed the yellow color found in the species of this section into the reservoir of *Camellia* cultivars. It is too early to say how successful this effort will be. Although F1 hybrids between *C. chrysantha* and both *C. reticulata* and *C. pitardii* var. *yunnanica* have been made at the Kunming Botanical Garden, these hybrids have so far proved sterile (Zhang Aoluo, per. com.). *C. chrysantha* is a diploid ($n=15$), this may account for the sterility problems when hybridization is attempted with the hexaploid ($n=45$) species *C. reticulata* and *C. pitardii* var. *yunnanica*. Greater success might be achieved by hybridizing *C. chrysantha* with diploid camellias such as some of the diploid cultivars of *C. japonica*. Because *C. chrysantha* is in subgenus *Thea* and most *Camellia* cultivars are in subgenus *Camellia* the phylogenetic distance may also present difficulties in breeding. There is, however, a yellow flowered species in subgenus *Camellia*, and although the flowers of *C. luteoflora* are very small, this species may prove to be a more important species for breeding yellow color into other camellias than the species in section *Chrysantha*.

Subgenus *Metacamellia* contains 57 species, but the small flowers characteristic of this group make them less desirable as cultivated plants. However, they should not be overlooked, as many species will no doubt make handsome shrubs and small trees in cultivation.

As *Camellia* is largely a subtropical genus, one of the main limitations for cultivating it in North America and Western Europe has been a lack of frost hardiness. This is not so much a problem in the coastal parts of California where even low elevation tropical species such as *C. chrysantha*, *C. granthamiana* and *C. hongkongensis* grow outside during the winter without much difficulty, but winter temperatures are certainly a limiting factor in Eastern United States and much of Western Europe. Figure 2 shows the distribution of species as well as the isotherms for the average January temperature in China between $-4°$ and $+12°$ C. These isotherms only give a general picture of temperatures faced by different species in China, but because no data are available on the frost hardiness of most *Camellia* species, these average minimum temperatures are useful as a guide to where different species might be able to grow in other parts of the world.

BRUCE BARTHOLOMEW

Department of Botany
California Academy of Sciences

LITERATURE CITED

Abel, C. 1818. Narrative of a journey in the interior of China, and of a voyage to and from that country, in the years 1816 and 1817. London.

Ackerman, W. L. 1978. Winter injury in the National Arboretum *Camellia* collection during the 1976–77 season. In D. L. Feathers and M. H. Brown, eds. The *Camellia;* its history, culture, genetics and a look into its future development. Columbia.

Aiton, W. 1789. Hortus Kewensis; or a catalogue of the plants cultivated in the Royal botanic garden at Kew. 3 vols. London.

———. 1810–1813. Hortus Kewensis; or a catalogue of the plants cultivated in the Royal botanic garden at Kew, ed. 2. 5 vols. London.

Andoh, Y. 1972. *Camellia reticulata* as illustrated in ancient books in Japan. New Zealand Camellia Bulletin 7: 4–20.

Andrews, H. C. 1797. *Camellia japonica* var. flore albo pleno. Andr. Bot. Repos. 1: pl. 25.

Anonymous. 1981. Species. Camellia Research News 1: 1–2.

———. 1982a. Abbreviations of English names of institutions cited in Acta Phytotaxonomica Sinica. Acta Phytotax. Sin. 20: 252–256.

———. 1982b. Species update. Camellia Research News 2: 5.

Bartholomew, B. 1981. A report on Hong-ta Chang's revision of the genus *Camellia*. American Camellia Yearbook 1981: 42–52.

———. 1982. *Tiannan Chahua Xiaozhi* and the old *Camellia reticulata* cultivars from Yunnan, China. American Camellia Yearbook 1982: 147–155.

Berlèse, L. 1837. Monographie du genre *Camellia,* ou, essai sur sa culture, sa description et sa classification. Paris.

———. 1841–1843. Iconographie du genre *Camellia* ou description et figures des camellias les plus beaux et les plus rares peints d'apres nature par J. J. Jung. 3 vols. Paris.

Booth, W. B. 1830. History and description of the species of *Camellia* and *Thea;* and the varieties of *Camellia japonica* that have been imported from China. Trans. Hort. Soc. London 7: 519–562.

Booth, W. B. and A. Chandler. 1831. Illustrations and descriptions of the plants which compose the natural order *Camellieae* and of the varieties of *Camellia japonica,* cultivated in the gardens of Great Britain. Vol. 1. London.

Bretschneider, E. 1880. Early European researches into the flora cf Cina. Journ. N. China Branch Roy. Asiat. Soc. n. ser. 15: 1–192.

———. 1898. History of European botanical discoveries in China. 2 vols. St. Petersburg.

Brown, M. H. 1978. History. In D. L. Feathers and M. H. Brown, eds. The *Camellia;* its history, culture, genetics and a look into its future development. Columbia.

Chang, H. T. 1981. A taxonomy of the genus *Camellia*. Acta Sci. Nat. Univ. Sunyatseni, monogr. ser. 1: 1–180.

Curtis, S. 1819. A monograph on the genus *Camellia* from original drawings by Clara Maria Pope. London.

Don, G. 1831–1838. A general history of the dichlamydeous plants comprising complete descriptions of the different orders, etc. 4 vols. London.

Edwards, G. 1747. The peacock pheasant of China. A natural history of uncommon birds and of some other rare and undescribed animals, etc. 2: pl. 67.

Feng, G. M, L. F. Xia and X. H. Zhu. 1981. Yunnan Shanchahua (Yunnan camellias). Kunming.

Henrey, B. 1976. British botanical and horticultural literature before 1800. 3 vols. London.

Hooker, W. J. 1858. *Camellia rosaeflora,* rose-flowered *Camellia.* Curtis's Bot. Mag. 84: pl. 5044.

Hu, H. H. 1938. Recent progress in botanical exploration in China. Journ. Roy. Hort. Soc. (London) 63: 381–389, 8 pls.

Hume, H. H. 1955. Camellias in America, ed. 2. Harrisburg.

Kaempfer, E. 1712. Amoenitatum exoticarum politico-physico-medicarum etc. Lemgoviae.

Kitamura, S. 1952. The longlived horticultural varieties of *Camelliae* in Kyoto, Japan. Acta Phytotax. Geobot. 14: 115–117.

_____. 1970. On taxonomical ranking of Camellia 'Uraku'. Acta Phytotax. Geobot. 24: 173–174.

Kondo, K. 1976. A historical review of taxonomic complexes of cultivated taxa of *Camellia.* American Camellia Yearbook 1976: 102–115.

Lemaire, C. 1869. 1° *Camellia japonica* L.; 2° *Camellia sasanqua* Thunb. Illustr. Hort. 16: pl. 581.

Linden, L. and E. Rodigas. 1886. Cronique horticole. Illustr. Hort. 33: 69–74.

Lindley, J. 1826. *Camellia euryoides, Eurya*-like *Camellia.* Bot. Reg. 12: pl. 983.

_____. 1827. *Camellia reticulata* Captain Rawes's Camellia. Bot. Reg. 13: pl. 1078.

_____. 1857. *Camellia reticulata;* flore pleno. Curtis's Bot. Mag. 83: pl. 4976.

Linnaeus, C. 1753. Species plantarum. 2 vols. Stockholm.

_____. 1766–1767. Systema naturae per regnal tria naturae secundum classes, ordines, genera, species cum characteribus differentiis synanymis, locis, ed. 12. 3 vols. Stockholm.

Loureiro, J. 1790. Flora Cochinchinensis etc. 2 vols. Ulyssipone.

Makino, T. 1910. Observations on the flora of Japan. Bot. Mag. Tokyo 24: 71–84.

_____. 1918. A contribution to the knowledge of the flora of Japan. Jour. Jap. Bot. 1: 39–42.

Nakai, T. 1940. A new classification of the Sino-Japanese genera and species which belong to the tribe *Camellieae* (II). Journ. Jap. Bot. 16: 691–708.

Parks, C. R. 1982. *Camellia crapnelliana* Tutcher. American Camellia Yearbook 1982: 143–146.

Parks, C. R., K. F. Case and K. R. Montgomery. 1967. A possible origin of anthocyanin (red) pigmentation in the flowers of *Camellia sasanqua.* American Camellia Yearbook 1967: 229–242.

Parks, C. R., K. Kondo and T. Swain. 1981. Phytochemical evidence for the genetic contamination of *Camellia sasanqua* Thunberg. Jap. Journ. Breed. 31: 167–182.

Petiver, J. 1702–1709. Gazophylacii naturae et artis decades X, etc. 2 vols. London.

Sealy, J. R. 1935. *Camellia reticulata*. Curtis's Bot. Mag. 158: pl. 9397.

———. 1937. Species of *Camellia* in cultivation. Journ. Roy Hort. Soc. (London) 62: 352–369.

———. 1958. A revision of the genus Camellia. London.

Sims, J. 1819. *Camellia sasanqua* Palmer's double sasanqua. Curtis's Bot. Mag. 46: pl. 2080.

Thunberg, C. P. 1784. Flora Japonica, etc. Lipsiae.

Uemoto, S., T. Tanaka and K. Fujieda. 1980. Cytogenetic studies on the origin of *Camellia vernalis*, I. On the meiotic chromosomes in some related *Camellia* forms in Hirado Island. Journ. Jap. Soc. Hort. Sci. 48: 475–482.

Wallich, N. 1820. An account of a new species of *Camellia* growing wild at Nepal. Asiat. Res. 13: 428–432.

Woodroof, W. E., ed. 1981. Camellia Nomenclature, ed. 17. San Pedro.

Yü, T. T. 1950. *Camellia reticulata* and its garden varieties. In P. M. Synge, ed. Camellias and Magnolias Conference Report. London.

———. 1964. The garden camellias of Yunnan. In F. Griffin, ed. Camellian. Columbia.

Yü, T. T. and B. Bartholomew. 1980. The origin and classification of the garden varieties of *Camellia reticulata*. American Camellia Yearbook 1980: 1–29.

Yü, T. T. and Y. Z. Feng. 1958. Yunnan Shanchahua Tuzhi (Illustrated account of *Camellia reticulata*). Beijing.

Chapter 1

Introduction

Camellia is the largest genus in Theaceae with over 200 species. There are several obvious differences between *Camellia* and the other 3 large genera in the family which are *Eurya* (140 spp.), *Ternstroemia* (130 spp.) and *Adinandra* (100 spp.). These differences are that: 1. The genus is more primitive; the flower parts are more numerous with less differentiation. 2. The distribution is concentrated and continuous. 3. *Camellia* contains very complex groups within the genus, yet the phylogenetic relations above the specific level are clearer than with the other genera in the family. 4. *Camellia* is economically very important, so it has been highly regarded.

In the 1920s there were only 40 recognized species (Melchior in Engler, Nat. Pflanzenfam. Aufl. 2, 21: 128–33. 1925), and in the 1950s this number had increased to 100 (Sealy, Rev. Gen. Camellia. 1958). Since Liberation, due to comprehensive and systematic investigations of the plant resources within China, there are now more than 200 recognized species subdivided into 4 subgenera and 20 sections, 90% of which are distributed in southern China, with 20 plus species distributed in the nearby tropical and subtropical areas. Among the 20 sections there are 2 sections with 3 species that do not occur in China. Of the 20 species that are scattered in nearby areas most of them are found in the northern part of the Indochina peninsula. There is only one species in the Philippines and Indonesia. Therefore, *Camellia* is a typical representative of the Cathaysian Flora.

TAXONOMY

Before the 1930s the botanical world was beginning to understand the genus *Camellia*. Melchior revised the genus and divided *Camellia* into 5 sections (Melchior in Engler, Nat. Pfflanzenfam. Aufl. 2, 21: 128-33. 1925). In Sealy's revision, 82 species were divided into 12 sections, but there were 20 species not placed in sections because not enough information was available to make their position clear (Sealy, Rev. Gen. Camellia. 1958). Although Sealy's excellent book is an important reference for the study of the genus *Camellia,* his work fell short in four ways: after dividing the genus into sections he did not emphasize the taxonomic relationships; some species were incorrectly placed in sections; he mixed up the species in some sections; and the specific diversity within the genus was not appreciated at the time Sealy wrote his book. Our work undertakes to rectify these weaknesses. The following are the taxonomic divisions used in the present work:

I. Subgen. *Protocamellia* Chang
  type *C. granthamiana* Sealy, 9 species.
  1. Sect. *Archecamellia* Sealy, emend. Chang
    type *C. granthamiana* Sealy, 3 species.
  2. Sect. *Stereocarpus* (Pierre) Sealy
    type *C. dormoyana* (Pierre ex Lanessan) Sealy, 5 species.

3. Sect. *Piquetia* (Pierre) Sealy
   type *C. piquetiana* (Pierre ex Lanessan) Sealy, 1 species.
II. Subgen. *Camellia*
   type *C. japonica* L., 73 species.
   4. Sect. *Oleifera* Chang
      type *C. oleifera* Abel, 4 species.
   5. Sect. *Furfuracea* Chang
      type *C. furfuracea* (Merr.) Cohen-Stuart, 8 species.
   6. Sect. *Paracamellia* Sealy
      type *C. kissii* Wall., 16 species.
   7. Sect. *Pseudocamellia* Sealy
      type *C. szechuanensis* Chi, 5 species.
   8. Sect. *Tuburculata* Chang
      type *C. tuburculata* Chien, 6 species.
   9. Sect. *Luteoflora* Chang
      type *C. luteoflora* Y. K. Li, 1 species.
   10. Sect. *Camellia*
      type *C. japonica* L., 33 species.
III. Subgen. *Thea* (L.) Chang
   type *C. sinensis* (L.) O. Kuntze, 61 species.
   11. Sect. *Corallina* Sealy
      type *C. corallina* Sealy, 12 species.
   12. Sect. *Brachyandra* Chang
      type *C. brachyandra* Chang, 11 species.
   13. Sect. *Longipedicellata* Chang
      type *C. longipedicellata* (Hu) Chang & Fang, 4 species.
   14. Sect. *Chrysantha* Chang
      type *C. chrysantha* (Hu) Tuyama, 10 species.
   15. Sect. *Calpandria* (Bl.) Cohen-Stuart
      type *C. lanceolata* (Bl.) Seem., 2 species.
   16. Sect. *Thea* (L.) Dyer
      type *C. sinensis* (L.) O. Kuntze, 18 species.
   17. Sect. *Longissima* Chang
      type *C. longissima* Chang & Liang, 2 species.
   18. Sect. *Glaberrima* Chang
      type *C. glaberrima* Chang, 2 species.
IV. Subgen. *Metacamellia* Chang
   type *C. cuspidata* (Kochs) Hort. ex Bean, 57 species.
   19. Sect. *Theopsis* Cohen-Stuart
      type *C. cuspidata* (Kochs) Hort. ex Bean, 43 species.
   20. Sect. *Eriandria* Cohen-Stuart
      type *C. caudata* Wall., 14 species.

Subgenus *Protocamellia* has 3 sections. The species in this subgenus have 5-locular ovaries and free styles. The perules are not differentiated into bracts and sepals and are comparatively large and persistent. The petals are relatively numerous. The stamens are mostly separate, reflecting the primitive position of the subgenus. The other three subgenera differ in all the above characteristics.

Subgenus *Camellia* is a large group. The perules are not differentiated into bracts and sepals; they increase in size from the outer to inner section of the flower and are deciduous after blooming. The perules and petals represent a transitional stage. The stamens are numerous, from free to connate, ovaries basically 3-locular, styles free to connate, capsules with a central columella. Each section within this subgenus is quite uniform and so providing the basic characteristic of the subgenus. The 7 sections within the subgenus are all natural groups showing common characteristics.

In the subgenus *Thea*, except for the two comparatively primitive sections whose perules are not completely differentiated, the other six sections possess characteristics opposite to subgenus *Protocamellia* and *Camellia*. The bracts and sepals are differentiated and basically persistent. The pedicels are usually present and sometimes unusually long. The petals are only slightly connate. The stamens are usually free with only a few connate into a filament tube. The capsules are tricoccus and with a columella. In this subgenus the number of bracts, their position, whether or not they are persistent, the length of the pedicel, whether or not the stamens are connate, and the length of the filament tube are all characteristics that are distinct for each section.

The perules of subgenus *Metacamellia* are differentiated into bracts and sepals and are persistent. In this respect it is the same as subgenus *Thea*. The unique characteristics are that the plants, leaves and flowers are much smaller; the sepals and petals are in 5s; the stamens are in 1-2 series; filaments are connate into a tube; the capsules are small; originally 3-locular ovaries become 1-locular at maturity and thus without the columella; and there is usually one seed.

The 3 sections of subgenus *Protocamellia* each has its own characteristics. Section *Archecamellia* is recognized by numerous, large, persistent, and undifferentiated perules. Section *Stereocarpus* has more petals; perules are smaller and differ from the previous section. In section *Piquetia* the flowers are united into an inflorescence, bracts and sepals are differentiated, and the stamens are free. These differences represent a further primary level of development.

The 7 sections of subgenus *Camellia* each has its own characteristics by which it can be recognized. Section *Camellia* has red-colored petals, highly connate stamens and styles. Section *Pseudocamellia* is differentiated from section *Camellia* by having white flowers and separate styles. The fruit of section *Tuberculata* has wrinkled skin and is differentiated from section *Pseudocamellia* by having pubescent seeds. In section *Furfuracea* the petals and stamens are only slightly connate, and the section is characterized by furfuraceous capsules. The flowers of section *Oleifera* are large and are differentiated from those of section *Furfuracea* by the petals and stamens being semi-free. Section *Oleifera* is separated from section *Paracamellia* by having more stamen series and a long style. Section *Luteoflora* is characterized by small yellow flowers with unpatent petals.

Of the 8 sections in subgenus *Thea*, two are primitive and differ from the others by having undifferentiated perules. Sections *Corallina* and *Brachyandra* can be separated based on the stamens and style length and whether or not the stamens and style are connate. Section *Chrysantha* has many persistent bracts and golden-yellow petals. Section *Longipedicellata* is differentiated by having even bracts, white petaled flowers and the stamens connate into a tube. Section *Thea* has 2 deciduous bracts and the petals and stamens nearly free. Section *Longissima*, like section *Thea*, has two deciduous bracts and free petals and stamens, but it has extremely long pedicels and capsules that dehisce from the bottom to the top. Sections *Glaberrima* and *Calpandria* have filaments formed into a tube. The former has 2 deciduous bracts so it is evolving along the same lines as section *Thea*. The latter has perules that are not differentiated and are persistent which shows similarities with section *Chrysantha*.

Subgenus *Metacamellia* has two closely related sections. They share many highly derived characteristics expressed in the flowers becoming smaller, the number of flower bracts and stamens decreasing, etc. Section *Theopsis* has glabrous ovaries, but section *Eriandria* has pubescent ovaries and anthers basifixed. This subgenus has some primitive members with 3-locular fruit and free stamens.

PHYLOGENY

The phylogeny of the genus *Camellia* can be traced by using the following rules.

1. The primitive condition is marked by undifferentiated perules which are numerous, spirally arranged and deciduous after blooming.

2. Numerous free petals mark a more primitive form than fewer connate petals.

3. Numerous free stamens is a primitive characteristic. Filaments becoming connate marks a derived characteristic, eventually evolving into a tube with no free filaments.

4. 5-locular ovaries and completely free styles is more primitive than 3-locular ovaries with joined styles.

5. The primitive condition is for the ovary to have several fertile carpels and a central columella. Conversely, a single fertile carpel and no columella is a derived characteristic.

Based on these observations, the evolutionary direction within the genus displays a reduction in the number of flower parts, a reduction of both the volume and linear dimensions of various flower and fruit structures, and changes in various flower structures from free to connate.

The phylogenetic development of the genus *Camellia* reflects obvious steps. Perules of the primitive types are undifferentiated, more numerous and spirally arranged. The species closest to the primitive genus *Camellia* have free stamens and 5 or more locular ovaries. Among the most primitive members of the genus *Camellia*, the apex of the ovary has 5 free styles which leave 5 isolated parts on the apex of the capsule. This phenomenon can be seen in *C. yunnanensis* and reflects the relation of Theaceae to the Dilliales. In the subgenus *Protocamellia*, specific populations are discontinuous which is also a primitive characteristic.

A separation into two distinct directions marks the second stage in the phylogeny of *Camellia*. In one group (subgenus *Camellia*) the perules remain undifferentiated and are deciduous after flowering. The flowers are sessile, and the 5-locules of the ovary are reduced to 3. The number of petals is reduced and the numerous stamens become connate. The style shows similar changes in all seven sections.

Subgenus *Thea* shows the other direction in which the genus has evolved with the perules differentiating into bracts and sepals. In its phylogenetic development, subgenus *Camellia* is closely related to section *Archecamellia*, but subgenus *Thea* is more closely related to section *Stereocarpus*, in that the bracts and sepals are persistent and not deciduous. As members of subgenus *Thea* evolved the trend has been for the bracts to decrease in number, the pedicel to lengthen and the various flower parts change from free to connate. These changes are characteristic for each section in subgenus *Thea*. The 8 sections are in turn divided into 3 groups. The most primitive group consists of sections *Corallina* and *Brachyandra* and is characterized by undifferentiated bracts. The second group is composed of sections *Longipedicellata*, *Calpandria* and *Chrysantha*, and this group has differentiated bracts which are numerous and persistent. The third group consists of sections *Longissima*, *Glaberrima* and *Thea*, and in this group the bracts are not only differentiated but decrease to two and are deciduous after blooming (Fig. 1).

Subgenera *Metacamellia* and *Thea* are directly related. Both have persistent bracts and sepals. Both probably evolved from section *Stereocarpus*, despite the fact that subgenus *Metacamellia* is characterized by reduction in the number of various flower parts and the fertility of the ovaries. The two sections within subgenus *Metacamellia* have a tendency towards parallel development. Their primitive representatives have free stamens and 3-locular fruit in addition to the change to basifixed anthers and connate

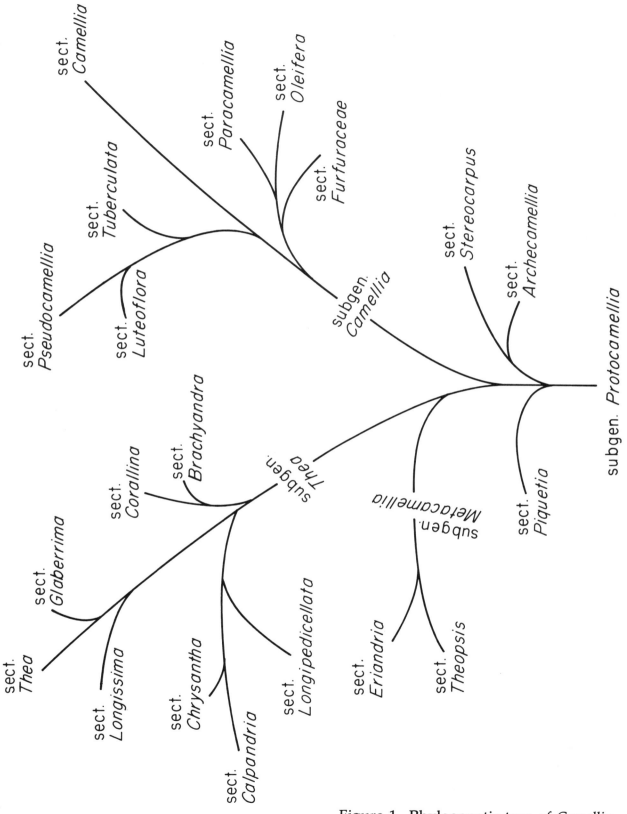

Figure 1. Phylogenetic tree of *Camellia*.

styles. The evolution of this subgenus is further exemplified by changes in the shape of the sepals and pubescent stamens and pistils.

The primitive representatives of subgenera *Camellia, Thea* and *Metacamellia* all have 5-locular ovaries which directly relates them to subgenus *Protocamellia*. The phylogenetic and distributional patterns are the same and clearly demonstrate their localized geographic origin.

There have been many interpretations of the taxonomic divisions within *Camellia*. Kaempfer, Blume, Rafinesque, Nees, Hallier, Nakai and Hu have divided *Camellia* into various genera, but these attempts have not been successful. By reason of inadequate information, these authors did not fully understand the phylogenetic development of *Camellia*. Based on the large number of specimens now known and a better developed understanding of the patterns of variation within the genus, it is possible to clearly define the three subgenera *Camellia, Thea* and *Metacamellia*.

Although these three subgenera are clearly defined, species in subgenus *Protocamellia* have some characteristics of all three of the other subgenera. For example, *C. granthamiana, C. albogigas* and *C. pleurocarpa* all have undifferentiated perules as is the case in subgenus *Camellia*, but the perules are persistent as in subgenera *Thea* and *Metacamellia*. As another example, in section *Stereocarpus, C. krempfii* has 15 undifferentiated and persistent perules, filaments connate into a tube and red petals. This species shares some characteristics with section *Camellia* of subgenus *Camellia* but also some with section *Chrysantha* of subgenus *Thea* as well as other characteristics with subgenus *Metacamellia*. It would be confusing to place *C. krempfii* into any of these three subgenera, but if this species is placed in a separate genus, the boundary lines between the genera are clouded. As a third example, *C. yunnanensis* was once established by H. H. Hu as the nomenclatural type of *Glyptocarpa,* but besides the 5 shallow troughs on the ovary apex, the other characteristics are not sufficiently different to separate it from the other members of subgenera *Protocamellia* or *Camellia*. These examples show that it is not possible to treat subgenus *Protocamellia* as a distinct genus. Were such a classification offered, the dividing lines between the genera would become confused and so destroy the natural system of the genus *Camellia* proposed here.

GEOGRAPHIC DISTRIBUTION

The genus *Camellia* contains about 200 species. Among woody dicots it cannot be compared with some of the large genera in Lauraceae, Fagaceae, Aquifoliaceae, Symplocaceae, Myrtaceae, Moraceae, Ericaceae, Rosaceae etc. which have more species. However, the area of distribution of *Camellia* is very concentrated, so the number of species per unit of area is comparable or even exceeds that of these larger genera. The center of distribution of *Camellia* is in south and southwestern China with its focus in the provinces of Yunnan, Guangxi and Guangdong straddling the Tropic of Cancer. Diversity decreases slowly towards the north and south (Fig. 2). In analyzing the distribution of the 20 sections one can see that most sections are concentrated along an east/west axis in this long and narrow region (Figs. 1 and 2). 8 sections are centered in Guangxi, 7 in Guangdong and 7 in Yunnan. The 3 sections in subgenus *Protocamellia* have 9 species scattered in Vietnam, Yunnan and Guangdong reflecting the historic significance of the region along the Tropic of Cancer as the center of distribution. Even now some primitive species are to be found only in this region. At present the center of distribution remains within the historic bounds so there has not been a significant shift from the original distribution.

Among the 20 sections many have their own special area of distribution within the historic central region. Section *Archecamellia* is centered in Guangdong and northern Vietnam. The center of distribution of the 5 species of section *Stereocarpus* lies between

Yunnan and northern Vietnam. The 4 species of section *Oleifera* are centered in Guangdong and Guangxi. Among the 8 species of section *Furfuracea*, 6 species are endemic to Guangdong, although some species in this section extend to Guangxi, Hunan, Jiangxi, Zhejiang and Fujian provinces. Section *Paracamellia* has the greatest distributional area. Its primary center is in Guangxi with secondary centers in Yunnan and Sichuan, while the section extends westward to Burma and India. 90% of the species in section *Pseudocamellia* are located in Sichuan, Guizhou and Yunnan provinces. Section *Tuberculata* shows the same distribution as section *Pseudocamellia*, which reflects their close relationship. Section *Camellia* has 33 species centered in Yunnan, Guangdong and Guangxi. Distribution of this section extends eastward to Zhejiang, Jiangxi and Fujian provinces which forms a secondary distributional center for subsection *Lucidissima*. In subgenus *Camellia* there are 4 sections in the southwestern mountainous parts of China, while the 2 remaining sections, *Furfuracea* and *Oleifera*, have their distributional centers in Guangdong and Guangxi provinces respectively.

Among the 8 sections in subgenus *Thea* the more primitive sections, *Corallina* and *Brachyandra*, have their centers of distribution in Yunnan and Guangxi provinces and northern Vietnam. This distribution is not an accident but reflects the phylogenetic position of these sections. The 2 species of section *Calpandria* depart from the center of distribution of this subgenus and reach the mountainous areas of Southeast Asia. The stamens connate into a tube reflects the more highly evolved character of this section. The 10 species of section *Chrysantha* are centered on a narrow belt of land astride the border of Guangxi Province and northern Vietnam. The center of distribution of section *Thea* is in Yunnan and Guangxi provinces. Interestingly, the distributional range of this large section is not extensive. The 2 remaining sections of this subgenus exhibit a marked degree of endemism and very restricted distribution.

The distribution of subgenus *Metacamellia* is equally concentrated and continuous indicating that it is a secondary and differentiated group. Except for a few species its distribution does not extend beyond southern China. Section *Theopsis* is the largest section in the genus, with its center of distribution from Yunnan to Guangdong provinces along the Tropic of Cancer in an east-west band. The eastern extreme extends to Taiwan and the western extreme terminates in the central part of Yunnan Province. The distribution of section *Eriandria* is narrow. The northern boundary does not go beyond 24°N latitude, while the southern boundary nearly reaches the border between China and northern Vietnam. This section extends westward to eastern India.

Not only because of the completeness of the phylogeny but also the concentrated distribution pattern, it is reasonable to suggest that south and southwestern China is both the modern and original center of *Camellia*. Within Theaceae, *Camellia* has more primitive characteristics than other genera and is the largest genus in the family. This is why we believe that Theaceae originated and developed in and radiated out from south China. Theaceae is one of the principle constituents of the Cathaysian flora. Geologically speaking this flora extends back to the Cretaceous period or earlier. Fossils of Theaceae including leaf-prints and fruit have been found recently in Tertiary-Cretaceous deposits from Chinglong County of southwestern Guizhou Province. Fossils of Theaceae are frequently found in Tertiary deposits of North America and Europe, and fossils of *Ternstroemioxylon* have been found in the Cretaceous strata of Egypt.

20 CAMELLIAS

Figure 2. The number of *Camellia* species in each geographic area and the isotherms within China for the average January temperature.

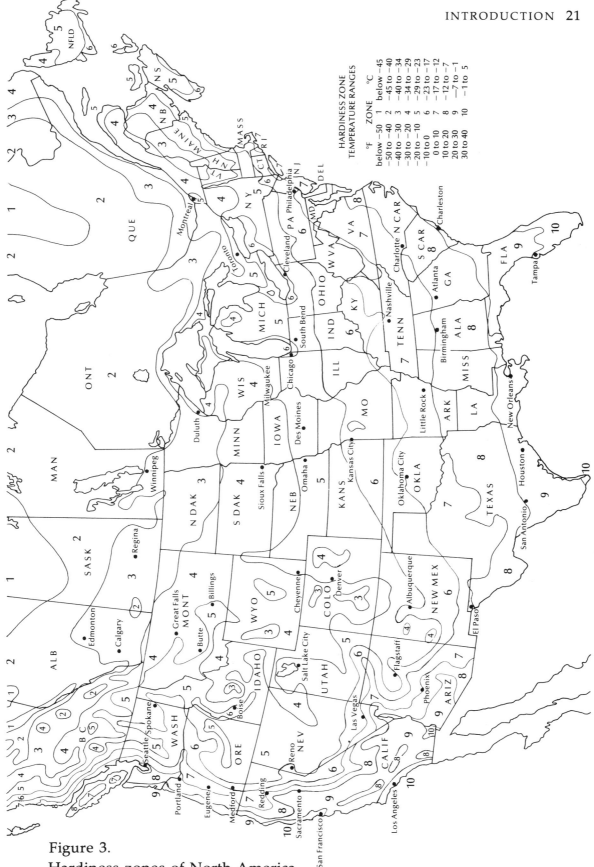

Figure 3.
Hardiness zones of North America.

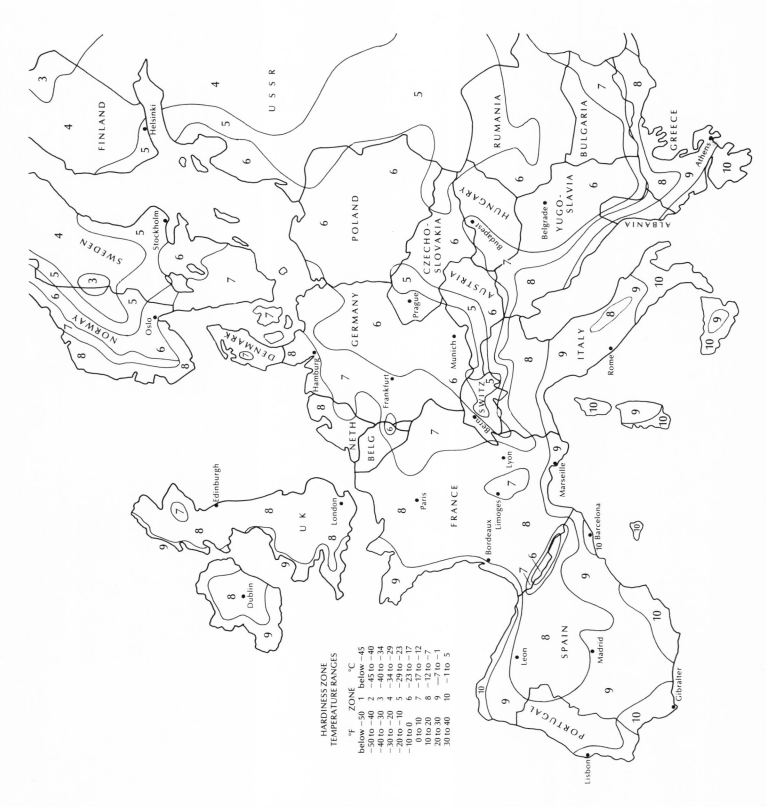

Figure 4. Hardiness zones of Europe.

TABLE 1. **The centers of distribution for each section and the number of species occurring in each area.**

	Total no. of species	Center of distribution
Subgen. *Protocamellia* Sect.		
Archecamellia	3	Guangdong (2), Vietnam (1)
Stereocarpus	5	Yunnan (3), Vietnam (2)
Piquetia	1	Vietnam (1)
Subgen. *Camellia* Sect.		
Oleifera	4	Guangxi (2), Guangdong (2), Japan (1)
Furfuracea	8	Guangdong (6), Guangxi (5)
Paracamellia	16	Guangxi (6), Yunnan (5)
Pseudocamellia	5	Sichuan (3), Yunnan (2)
Tuberculata	6	Guizhou (3), Sichuan (2), Hunan (2)
Luteoflora	1	Guizhou (1)
Camellia	33	Guangxi (12), Yunnan (8), Guangdong (7) Hunan (7)
Subgen. *Thea* Sect.		
Corallina	11	Yunnan (3), Vietnam (2), Guangxi (2)
Brachyandra	12	Yunnan (5), Hainan (4), Vietnam (3)
Longipedicellata	4	Vietnam (3), Guangxi (2)
Chrysantha	10	Guangxi (8), Vietnam (4)
Calpandria	2	Thailand (1), Philippines and Indonesia (1)
Thea	18	Yunnan (13), Guangxi (10)
Longissima	2	Guangxi (1), Vietnam (1)
Glaberrima	2	Guangdong (1), Yunnan (1)
Subgen. *Metacamellia* Sect.		
Theopsis	43	Guangxi (14), Guangdong (12), Yunnan (11)
Eriandria	14	Guangdong (9), Guangxi (5)

TABLE 2. **The geographic distribution of *Camellia***

Area	No. of species	No. of endemic species	No. of sections	No. of subgenera
Guangxi	68	25	12	3
Yunnan	57	29	13	4
Guangdong (excl. Hainan)	47	17	10	4
Sichuan	30	8	10	4
Vietnam	26	16	15	4
Guizhou	24	7	11	3
Hunan	24	3	8	3
Jiangxi	24	2	7	3

Area	No. of species	No. of endemic species	No. of sections	No. of subgenera
Fujian	13	0	7	3
Hainan	12	5	7	3
Zhejiang	12	1	7	3
Hubei	10	0	4	2
Taiwan	8	2	4	2
Burma	8	0	5	3
Anhui	6	0	5	3
Jiangsu	6	0	4	3
Laos	5	0	4	2
Japan	4	3	4	2
India	4	1	2	2
Thailand	2	1	2	2
Shaanxi	2	0	2	2
Xizang (Tibet)	2	0	2	2
Bhutan	2	0	2	2
Korea	1	0	1	1
Shandong	1	0	1	1
Sikkim	1	0	1	1
Nepal	1	0	1	1
Cambodia	1	0	1	1
North Borneo	1	0	1	1
Indonesia	1	0	1	1
Philippines	1	0	1	1

ECONOMIC USE

The genus *Camellia* is of great economic value particularly as one species is the source of the extremely important beverage, tea. Tea has a long history of cultivation and many cultivars. Tea is made from the young leaves of *C. sinensis* var. *sinensis* and *C. sinensis* var. *assamica*.

Section *Thea* contains 18 species, but only one species is exploited to any extent. Leaves of some of the other species are collected and used locally but are not of major economic importance. For example, the famous *C. ptilophylla* of Longmen in Guangdong Province is locally known as *Baimao* (white hair) Tea. It has been used for centuries as a beverage, but it is not known outside the immediate area of Longmen. The economic potential for tea production from additional species is very great if underutilized species are brought into production.

There is a long history of the use of tea by the Chinese people. Based on textual evidence, tea was used as a medicine in the Shang Dynasty, 2,700 years B.C. After the Zhou Dynasty (ca. 1,000 B.C.) it was used as a beverage. The *Erya*, a Chinese classical dictionary, dating from some time between the Spring and Autumn Period to Western Han (ca. 600–300 B.C.), mentions tea as *Jia*. In the later herbal *Tangbencao* (669 A.D.) the word *Cha* is used. Thus, domestication and use as well as geographic distribution lead to the conclusion that tea is a Chinese endemic. However, J. Hutchinson has written that tea originated in India (Hutchinson, The Families of Flowering Plants, 3rd ed., 334. 1973). He theorized, without supporting evidence, that in the long distant past, Chinese entered India and brought back seeds to cultivate in China. This subjective conjecture is not only without any basis in fact, but is controverted by modern evidence. At present throughout south China wild tea is native in the evergreen forests. The mid-elevation evergreen forests of Bawangling on Hainan Island contain wild tea plants 15 m tall with trunk diameters of 40 cm. Wild tea is also found in the evergreen forests of Tianqing Shan, Ruyuan in Guangdong Province. Wild tea plants in Nannuo Shan, Menghai of Yunnan Province are up to 20 m tall with trunk diameters of 70 cm. These wild teas have large leaves and are *C. sinensis* var. *assamica*. It is an unfortunate accident of history that the name *C. sinensis* var. *sinensis* must be used for the more commonly cultivated plant and *C. sinensis* var. *assamica* for the wild species.

In the Tang Dynasty (7th century A.D.) tea was introduced to Japan. During the 18th century the European colonists introduced tea into various plantations in Southeast Asia. According to a report from the East India Company of England, wild tea was not found in India. It could be that around the 17th century the native *Dan* minority of Assam brought tea back from China and started plantations. In 1826 a member of the East India Company collected the species from a plantation. It was named *C. ? scottiana* by Wallich and entered into his Catalogue. In 1844 G. J. Gordon introduced tea seeds from China to Calcutta. The seedlings were given to Dehra Dun and Kumaoan for cultivation. From these origins the cultivation of tea began in India.

Tea oil is an important cooking oil in China's southern provinces. In various southern provinces programs of selective breeding and spreading of important cultivars are now underway. As an oil crop, among the 4 species of section *Oleifera*, *C. oleifera* has the longest history of cultivation, and *C. gauchowensis* and *C. vietnamensis* are beginning to be utilized. *C. semiserrata*, *C. chekiangoleosa*, *C. reticulata* etc. among the 33 species within section *Camellia* are also very valuable oil crops. The many species of section *Furfuracea* and *Paracamellia* are beginning to be introduced as oil crops. These include *C. crapnelliana* and *C. grijsii*. In summary, the seeds of the 200 species of *Camellia* all contain oil, and many species provide useful oil crops to varying degrees. The oil is used both for cooking oil and for industrial purposes.

The pericarp of the *Camellia* capsules contains tannic acid. It is used both in adhesives and to increase coagulation of concrete.

The leaves of *Camellia* contain Xanthin, Theophylin, Theobromin, Adenin etc. in addition to Theanine, Glycoside, oleic acid, esters, etc. all of which are important ingredients for the pharmaceutical industry. People often use *C. japonica* for haemorrhage. The roots of *C. oleifera* are used to treat broken bones and burns, and *C. chrysantha* and *C. longipedicellata* are used to treat dysentery.

Camellia species are of great ornamental value, especially those species of section *Camellia*, which have been developed through a long history of selective breeding and cultivation by the workers of China. These cultivars are renowned internationally. Recently, the discovery and introduction of *C. chrysantha* has added a new species to horticulture in China and has received the attention of *Camellia* horticultural societies worldwide.

Following the flourishing of socialism in China, the natural resource of *Camellia* plants will be increasingly utilized and mobilized for the Four Modernizations and to realize the latent potentials of this remarkable genus to promote the livelihood and welfare of the people.

Chapter 2

Camellia

Camellia L., Sp. Pl. 2: 698. 1753.
  Thea L., Sp. Pl. 1: 515. 1753.
  Tsubaki Kaempfer ex Adanson, Fam. Pl. 2: 399. 1763.
  Tsia Kaempfer ex Adanson, Fam. Pl. 2: 459, 613. 1763.
  Calpandria Bl., Bijdr. Fl. Nederland. Indië 1: 178. 1825.
  Theaphylla Raf., Med. Fl. 2: 267. 1830.
  Sasanqua Nees in Siebold, Nippon 2: 13. 1832.
  Kemelia Raf., Sylva Tellur., 139. 1838.
  Demitus Raf., Sylva Tellur., 139. 1838.
  Drupifera Raf., Sylva Tellur., 140. 1838.
  Piquetia (Pierre) H. Hallier, Beih. Bot. Centralbl. 39(2): 162. 1921.
  Stereocarpus (Pierre) H. Hallier, Beih. Bot. Centralbl. 39(2): 12. 1921.
  Camelliastrum Nakai, Journ. Jap. Bot. 16: 699. 1940.
  Theopsis (Cohen-Stuart) Nakai, Journ. Jap. Bot. 16: 704. 1940.
  Yunnanea Hu, Acta Phytotax. Sin. 5: 282. 1956.
  Glyptocarpa Hu, Acta Phytotax. Sin. 10: 25. 1965.
  Kailosocarpus Hu, Scientia 1957: 170. 1957, *nom. nud.*
  Parapiquetia Hu, Scientia 1957: 170. 1957, *nom. nud.*

  Shrubs or trees. Leaves coriaceous, pinnately veined, often serrate, petiolate, rarely sessile and amplexicaul. Flowers hermaphroditic, solitary or in clusters at the branch terminals or leaf axils, pedicellate or sessile; bracts usually 2-8, sepals usually 5-6, sometimes the differentiation between bracts and sepals is indistinct, becoming perulate, to 21 perules, deciduous or persistent; corollas white, red or yellow; petals 5-12, basally connate; stamens numerous, in 2-6 series, outer filament whorl often connate into a filament tube, adnate to petal bases; anthers dorsifixed or occasionally basifixed, 2-locular, longitudinally cleft; ovaries superior, 3-5-locular, sometimes unilocular, 3-5-valvate, usually dehiscent from the top; columella persistent or lacking; seeds globose or polygonal, seed coats corneus, endosperm with high oil content.
  Ca. 200 species; ca. 180 in southern China, the remainder in the Indochina peninsula, eastern India, Malaysia, Japan and Philippines.
  Type: *C. japonica* L.

KEY TO SUBGENERA AND SECTIONS

1. Ovaries 5-locular; styles 5-parted, free or rarely connate; bracts and sepals not differentiated or not distinctly differentiated, usually 12-18, rarely 8, persistent; pedicellate, rarely sessile..................... I. Subgen *Protocamellia* Chang
  2. Flowers solitary, leaves 10-31 cm long.
    3. Bracts and sepals 12-16, large, 2-5 cm long, enclosing the capsule ....
      .... (I). Sect. *Archecamellia* Sealy

   3. Bracts and sepals 6-13, small, 0.3-1.5 cm long................................
      .... (II). Sect. *Stereocarpus* (Pierre) Sealy
  2. Flowers several on leafless short branches, leaves 29-42 cm long............
      .... (III). Sect. *Piquetia* (Pierre) Sealy
1. Ovaries usually 3-locular, styles 3-parted or 3-cleft; ovaries seldom 4-5-locular, but a single style, apically 4-5-cleft, rarely 4-5 free; bracts and sepals differentiated or not, 4-12, seldom more, persistent or deciduous.
  4. Bracts and sepals undifferentiated, more than 10, deciduous after anthesis; flowers usually large, 5-10 cm in diameter (2-4 cm in sect. *Paracamellia*), sessile; ovaries usually 3-locular, rarely 4-5-locular.............. II. Subgen. *Camellia*
    5. Filaments free, or basally slightly connate, not forming a tube; petals free or slightly connate, white.
      6. Flowers relatively large, 5-10 cm in diameter; stamens 1-1.8 cm long; capsules relatively large, 3-5-locular; styles ca. 1-1.5 cm long.
        7. Perules coriaceous, stamens in 3-5 series, styles connate, capsules not furfuraceous, pericarp woody or coriaceous... (IV). Sect. *Oleifera* Chang
        7. Perules easily deciduous, stamens in 2-3 series, styles free, capsules furfuraceous; pericarp loose, brown...... (V). Sect. *Furfuracea* Chang
      6. Flowers small, petals nearly free, stamens and styles 0.2-0.8 cm long; capsules small, often 1-locular, 1-2 cm in diameter, not furfuraceous
        .... (VI). Sect. *Paracamellia* Sealy
    5. Filaments connate into a short tube; petals basally highly connate.
      8. Flowers white or yellow; styles 3(-5)-parted, free; sepals membranous, deciduous or semi-persistent.
        9. Flowers white, fully patent.
          10. Ovaries not grooved, capsules smooth, seeds glabrous, petals broadly obovate................ (VII). Sect. *Pseudocamellia* Sealy
          10. Ovaries grooved, capsule surface tuberculate, seeds often pubescent, petals oblong.............. (VIII). Sect. *Tuberculata* Chang
        9. Flowers yellow, not patent.............. (IX). Sect. *Luteoflora* Chang
      8. Flowers red, sometimes almost white; styles connate, apex shallowly 3(-5)-cleft... (X). Sect. *Camellia*
  4. Bracts and sepals usually clearly differentiated, occasionally not differentiated and persistent, sepals persistent, bracts persistent or deciduous; flowers relatively small, 2-5 cm in diameter, pedicellate; stamens free, rarely with a filament tube; ovaries and capsules 3(-5)-locular, rarely 1-locular.
    11. Ovaries 3(-5)-locular, each locule equally fertile; fruit large, pericarp relatively thick, with a columella, sepals persistent, bracts persistent or deciduous, styles 3(-5)-parted or 3(-5)-cleft .... III. Subgen. *Thea* (L.) Chang
      12. Bracts not completely differentiated from sepals, persistent; pedicels extremely short, seldom 1 cm long.
        13. Filament bases free or slightly connate.
          14. Filaments linear, free, 1 cm long; styles 3-parted, rarely connate, 6-10 mm long..................... (XI). Sect. *Corallina* Sealy
          14. Filaments relatively thick, somewhat connate, 3-6 mm long; styles connate or free, 1-4 mm long........................
            .... (XII). Sect. *Brachyandra* Chang
        13. Filaments completely connate into a long tube, anthers born on the tube wall; styles short, apically 3-cleft..........................
          .... (XV). Sect. *Calpandria* (Bl.) Cohen-Stuart
      12. Bracts clearly differentiated from sepals, persistent or deciduous, sepals persistent; pedicels 6-30 mm long, usually 10 mm long.
        15. Bracts 5-11, persistent.
          16. Filaments free or basally slightly connate, flowers golden yellow;

              styles 3-parted, free .... (XIV). Sect. *Chrysantha* Chang
       16. Filaments with a short tube, flowers white, styles 3-parted or 3-cleft..................... (XIII). Sect. *Longipedicellata* Chang
  15. Bracts 2, early deciduous.
      17. Filaments nearly free.
          18. Capsules dehiscent from the apex, pedicels 5-6 mm long .... (XVI). Sect. *Thea* (L.) Dyer
          18. Capsules dehiscent from the base, pedicels ca. 30 mm long..................... (XVII). Sect. *Longissima* Chang
      17. Filaments basally connate into a short tube, glabrous... (XVIII). Sect. *Glaberrima* Chang
11. Ovaries usually with only 1 fertile locule; fruit small, wall shell thin, without a columella; bracts and sepals persistent, stamens in 1-2 series; styles long, connate, apically 3(-5)-cleft.............. IV. Subgen. *Metacamellia* Chang
    19. Filaments free or basal half connate, glabrous or pubescent; ovaries glabrous, anthers dorsifixed .... (XIX). Sect. *Theopsis* Cohen-Stuart
    19. Filaments extremely connate, rarely free, usually pubescent; ovaries pubescent, capsules pubescent, anthers basifixed.................... .... (XX). Sect. *Eriandria* Cohen-Stuart

1 *Camellia granthamiana* Sealy
   1. flowering branch;
   2. pistil.

Chapter 3

SUBGENUS *Protocamellia*

SUBGEN. I.
Protocamellia Chang, Acta Sci. Nat. Univ. Sunyatseni, monogr. ser. 1: 15. 1981.

Floribus terminalibus solitariis vel rarius racemosis, pedicellis brevibus, perianthiis pluribus 18-25, bracteis sepalisque indistinctis valde magnis peristentibus, staminibus multo-seriatis liberis vel rarius connatis ovario 5-locularii, stylis liberis 5 vel rarius connatis. Capsulis 5-valvatis columnaris.

Flowers terminal, solitary or several born on leafless branchlets, pedicels short; bracts, sepals and petals 18-25; bracts not differentiated from sepals, perules large, orbicular, persistent; stamens in numerous series, free, or occasionally connate; ovaries 5-locular; styles 5-parted, free, occasionally styles single. Capsules 5-locular, 5-valvate dehiscent, columella persistent.

3 sections, 9 species.
Type: *C. granthamiana* Sealy

SECTION I.
Archecamellia Sealy, Rev. Gen. Camellia, 36. 1958; Chang, Acta Sci. Nat. Univ. Sunyatseni, monogr. ser. 1: 15. 1981, *diagnosis emend.*

Floribus terminalibus solitariis, bracteis sepalisque tenue coriaceis, 12-16 magnis 2-4 cm longis, ovario 5-locularii, stylis 5 liberis.

Flowers solitary, terminal; bracts and sepals 12-16, thinly coriaceous, large, 2-4 cm long, persistent, enclosing the fruit; ovaries 5-locular; styles 5-parted, free.

3 species; 2 in China, 1 in Indochina.
Type: *C. granthamiana* Sealy

KEY TO SECT. *ARCHECAMELLIA*

1. Perules gradually increasing in size from outer to inner, overlapping; styles 2 cm long, fruit smooth.
   2. Perules 12, largest 3-4.5 cm long; stamens connate for 6 mm, styles shallowly 5-cleft .... 1. *C. granthamiana* Sealy
   2. Perules 17, largest 2.5 cm long; stamens nearly free, styles divided for more than one-half their length .... 2. *C. albogigas* Hu
1. Perules 16, lower 8 minute, born on the pedicel, largest 2.5 cm long; styles 6 mm long, fruit ribbed .... 3. *C. pleurocarpa* (Gagnep.) Sealy

1. *Camellia granthamiana* Sealy, Journ. Roy. Hort. Soc. 81: 182. 1956.
   Flowers large, 10-14 cm in diameter, white; bracts and sepals 12, to 5 cm long, completely enclosing the capsule; petals 8, stamens connate for 5-6 mm, styles apically 5-parted. Capsules 5-valvate dehiscent, seeds contain oil that could be of economic use.

Hong Kong: Zhang Hongda (H. T. Chang) 6520; Damao Shan, Zhang Hongda (H. T. Chang) 6562.

Guangdong: Lufeng, Bawan, Luojingchang Forest Center, Wei Zhaofen (C. F. Wai) 121328; Dapu, Antian, Wanmu Forest Center, Geobot. Lab. SCBI 7933.

2. *Camellia albogigas* Hu, Acta Phytotax. Sin. 10: 132. 1965.

This species is very close to *C. granthamiana*, but there are up to 17 bracts and sepals which are also smaller, the fruit does not completely enclose the capsule; petals 10, filaments connate into a short tube or nearly free; styles 5-cleft to nearly one-half the length, lower half connate; 1 seed per locule of the capsule.

Guangdong: Fengkai Xian, Chixing Commune, Qiu Huaxing 182; same loc., Huanggang Shan, Chen Shaoqing (S. C. Chun) 18405 (holotype in SCBI).

3. *Camellia pleurocarpa* (Gagnep.) Sealy, Rev. Gen. Camellia, 38. 1958.
   Thea pleurocarpa Gagnep., Not. Syst. 10: 130. 1942.

Branchlets glabrous. Leaves elliptic, 11-15 cm long, 5-6.5 cm wide, bases cuneate, glabrous. Capsules with 1.5 cm long pedicels, oblate, 3.5 cm wide, 16 persistent sepals subtending a 5-locular capsule, combined diameter 5 cm; styles 5-parted, free.

Distribution: Vietnam.

SECTION II.
Stereocarpus (Pierre) Sealy, Rev. Gen. Camellia, 45. 1958.

Thea sect. *Stereocarpus* Pierre, Fl. For. Cochinchine 2: sub. pl. 119. 1887.

Flowers terminal, solitary, nearly sessile; bracts and sepals 8-13, persistent; petals 10-13, stamens free or connate, ovaries 5-locular, styles free or connate.

5 species; 3 in China, 2 in Vietnam.

Type: *C. dormoyana* (Pierre ex Lanessan) Sealy

This section is distinct from sect. *Archecamellia* in that the perules are fewer in number and smaller, and all species have shorter petals.

KEY TO SECT. *STEREOCARPUS*

1. Leaves 30 cm long, bases cordate; flowers red, with an obvious pedicel; bracts and sepals 12-13, ovaries pubescent or glabrous. . . . . . . 1. *C. krempfii* (Gagnep.) Sealy
1. Leaves shorter, bases not cordate, flowers white, pedicels extremely short, bracts and sepals 8-10, ovaries glabrous.
    2. Leaves 11-18 cm long; styles connate, apices shallowly 5-cleft . . . . 2. *C. dormoyana* (Pierre ex Lanessan) Sealy
    2. Leaves 4-9 cm long; styles free or apical half free, 5-parted.
        3. Styles completely free, ovaries glabrous, leaf bases rounded.
            4. Leaves elliptic, 4-7 cm long, apices acute; pericarp 5-8 mm thick . . . . 3. *C. yunnanensis* (Pitard ex Diels) Cohen-Stuart
            4. Leaves lanceolate, 6-9 cm long, apices caudate-acuminate; pericarp 1-1.5 mm thick . . . . 4. *C. liberistyla* Chang ex Chang
        3. Lower half of styles connate, ovaries pubescent; leaves elliptic, bases cuneate . . . . 5. *C. liberistyloides* Chang

1. *Camellia krempfii* (Gagnep.) Sealy, Kew Bull. 1949(2): 219. 1949.
   Thea krempfii Gagnep., Not. Syst. 10: 127. 1942.

Branchlets glabrous. Leaves oblong, 30 cm long, subsessile, bases cordate, lateral veins to 20 pairs, petioles 1-1.5 cm long. Bracts and sepals 15, persistent; petals 10, 3.5

Camellia albogigas Hu
1. flowering branch; 2. perules; 3. stamens; 4. pistil; 5. seed; 6 capsule.

cm long, red; filaments basally connate into a tube; ovaries 5-locular, pubescent; styles 5-parted, free. This species is close to subgen. *Camellia* sect. *Camellia*. It is a transitional species to this section.

Vietnam: Kuizhou, Kecan Tung Huona Region, elev. 170-210 m, tree 4 m tall, evergreen forest, 25 January 1965, Exped. Sinica-Vietnamica 1790.

2. *Camellia dormoyana* (Pierre ex Lanessan) Sealy, Rev. Gen. Camellia, 45. 1958.
Thea dormoyana Pierre ex Lanessan, Pl. Util. Colon. Franc., 296. 1886.
Stereocarpus dormoyanus (Pierre ex Lanessan) H. Hallier, Beih. Bot. Centralbl. 39(2): 162. 1921.

Branchlets glabrous. Leaves ovate-oblong, 20 cm long, 8 cm wide, glabrous. Pedicels extremely short, bracts 2-4, sepals 5-6, petals 12, filament bases slightly connate; ovaries 5-locular, glabrous; styles connate, apices 5-cleft.

Vietnam: M. Poilane 19809, 21998.

Distribution: Vietnam.

3. *Camellia yunnanensis* (Pitard ex Diels) Cohen-Stuart, Meded. Proefst. Thee 40: 68. 1916.
Thea yunnanensis Pitard ex Diels, Notes Roy. Bot. Gard. Edinb. 5: 284. 1912.
Pyrenaria camellioides Hu, Bull. Fan Mem. Inst. Bio. Bot. 8: 136. 1938.
Kailosocarpus camellioides Hu, Scientia 1957: 170. 1957, *nom. nud.*
Glyptocarpa camellioides (Hu) Hu, Acta Phytotax. Sin. 10: 25. 1965.

Branchlets pubescent. Leaves oblong, 5-8 cm long, pubescent. Flowers sessile, bracts and sepals 10, petals 10-12, filaments free; ovaries 5-locular, glabrous or puberulent; styles 5-parted, free. Capsules 5-locular, 1-2 seeds per locule, valves 6-8 mm thick.

Yunnan: Wuding, Shilada, Xin Jingsan (G. S. Sin) 104; Chen Mou (M. Chen) 2203; Dali, Wang Hanchen (H. C. Wang) 4081; Weishan Xian, Jiang Ying (Y. Tsiang) 11889; Zhenkang, Yu Dejun (T. T. Yü) 17071; Feng Guomei (K. M. Feng) 21614, 22258; Liu Shenno (S. N. Liu) 21724; Zhenkang, Wang Qiwu (C. W. Wang) 72468 (holotype of *Pyrenaria camellioides* in PE).

Sichuan: Miyi, Wu Sugong (S. K. Wu) 64; Yanbian, Wu Sugong (S. K. Wu) 453, 587; Huili, Yu Dejun (T. T. Yü) 1611.

The capsule apices are flattened. In the process of fruit development an apical depression sometimes becomes torn. This is not a consistent shape. Of the specimens mentioned above, Feng 21614 does not have this apical depression, but Wang 72478 and Jiang 11898 do have the depression.

4. *Camellia liberistyla* Chang ex Chang, *sp. nov.*; Chang, Acta Sci. Nat. Univ. Sunyatseni, monogr. ser. 1: 18. 1981, *nom. invalid.*

Species *C. yunnanensem* proxima, sed foliis lanceolatis 6-9 cm longis, 1.5-2.3 cm latis, apice caudato-acuminatis, ovario glabro, pericarpio valde tenui 1-1.5 mm crasso differt.

Arbor parva, ramulis pubescentibus. Folia lanceolata 6-9 cm longa 1.5-2.3 cm lata, apice caudato-acuminata basi rotundata, subtus ad costam pubescentia, nervis lateralibus 8-9-jugis, margine serrulata, petiolis 3-5 mm longis. Flores terminales sessiles; bracteis et sepalis 9-10 scariosis suborbicularibus 6-12 mm longis sparse pubescentibus; petalis 10 obovatis 2 cm longis basi connatis; staminibus liberis petalis brevioribus, ovariis glabris 5-locularis, stylis 5 liberis. Capsulae subglobosae 3-3.5 cm diametro 5-valvatae dehiscentes valvis 1-1.5 mm crassis.

Small tree, branchlets pubescent. Leaves coriaceous, lanceolate, 6-9 cm long, 1.5-2.3 cm wide, apices caudate-acuminate, bases rounded, slightly shiny above, midveins pubescent below, lateral veins 8-9 pairs, margins serrulate, petioles 3-5 mm long. Flowers terminal, sessile; bracts and sepals 9-10, scarious, suborbicular, 6-12 mm long, scarcely pubescent; petals 10, obovate, 2 cm long; stamens free; ovaries glabrous, 5-locular; styles 5-parted, free. Capsules subglobose, 3-3.5 cm in diameter, apices with

Camellia yunnanensis (Pitard ex Diels) Cohen-Stuart flowering branch.

free styles or when ripe often torn off, 5-valvate dehiscent, valves 1-1.5 mm thick.

Yunnan: Luquan, Mao Pinyi (P. I. Mao) 1480 (*holotypus* in KUN); Wen Shan, elev. 1900-2300 m, Feng Guomei (K. M. Feng) 11264; Luquan, Mao Yinpin (P. I. Mao) 1537; Xinpin Xian, Wu Sugong (S. K. Wu) 436; Yan Shan 11570, Wang Qiwu (C. W. Wang) 85044.

5. *Camellia liberistyloides* Chang, Acta Sci. Nat. Univ. Sunyatseni, monogr. ser. 1: 20. 1981.

A *C. liberistyla* foliis oblongis basi cuneatis apice breviter acutis, staminibus connatis, ovario pubescenti, stylis inferioribus dimidiis connatis pubescentibus differt.

Frutex vel arbor parva, ramulis pubescentibus. Folia oblonga vel elliptica 3.5-5.5 cm longa 1.7-2.3 cm lata, apice acuta vel interdum obtusa et acuminata, basi late cuneata, supra in sicco nitidula, subtus ad costam saltem pubescentia, nervis lateralibus 6-8-jugis, margine serrulata, petiolis 3-5 mm longis plus minusve pubescentibus. Flores axillares albi sessile, bracteis et sepalis circ. 9, maximis suborbicularibus 8-9 mm longis extus sericeis; petalis 8-9 basi leviter connatis obovatis 1.3-1.8 cm longis; staminibus multo-seriatis 1.2-1.4 cm longis, filamentis extimis dimidiis connatis; ovariis pilosis 5-locularibus, stylis 1 cm longis 5-lobatis, lobis 5 mm longis, inferioribus pubescentibus.

Shrub or small tree, branchlets pubescent. Leaves oblong to elliptic, 3.5-5.5 cm long, 1.7-2.3 cm wide; apices acute, sometimes acuminate or slightly obtuse; bases broadly cuneate; slightly shiny above in the dry state; yellowish-brown below, midvein pubescent; lateral veins 6-8 pairs, margins serrulate; petioles 3-5 mm long, somewhat pubescent. Flowers terminal or axillary, white, sessile; bracts and sepals ca. 9, largest suborbicular, 8-9 mm long, exterior sericeous; petals 8-9, slightly connate at base, obovoid, 1.3-1.8 cm long; stamens in many series, outer whorl basally connate, 1.2-1.4 cm long; ovaries pubescent, 5-locular; styles 5-parted, 1 cm long, basal half connate, pubescent, upper half glabrous.

Yunnan: without loc., 21 December 1940, Liu Shenno (S. N. Liu) 17845 (holotype in PE).

SECTION III.
Piquetia (Pierre) Sealy, Rev. Gen. Camellia, 108. 1958.

Thea sect. *Piquetia* Pierre, Fl. For. Cochinchine 2: sub. pl. 119. 1887.

Leaves 30-40 cm long. Flowers solitary or in groups of 3-5 on short branches, pedicellate; bracts 2-3, persistent; sepals 5, persistent; petals 8 or more, stamens free, ovaries 5-locular, styles 5(-6)-parted.

1 species in Vietnam.

Type: *C. piquetiana* (Pierre ex Lanessan) Sealy

1. *Camellia piquetiana* (Pierre ex Lanessan) Sealy, Rev. Gen. Camellia, 108. 1958.
Thea piquetiana Pierre ex Lanessan, Pl. Util. Colon. Franc., 296. 1886.
Piquetia piquetiana (Pierre ex Lanessan) H. Hallier, Beih. Bot. Centralbl. 39(2): 262. 1921.

Branchlets glabrous. Leaf blades oblong, 30-42 cm long, 10-12.5 cm wide, glabrous, petioles 1 cm long. Flowers light purple, 4-5 cm in diameter. For the other characteristics see the section description.

Distribution: Vietnam.

Chapter 4

SUBGENUS *Camellia*

SUBGEN. II.
Camellia

Flores 1-2 terminales subsessiles, bracteis sepalisque indistinctis 8-21 involucratis 2.5-4 cm altis deciduis, petalis 6-12 liberis vel connatis, ovario 3-locularii vel rarius 5, stylis 3 liberis vel apice 3-fidis, capsulis 3-locularibus, columnaris, seminibus 1-5 in quoque loculo.

Flowers 1-2 terminal, subsessile; perules not differentiated into bracts and sepals, gradually becoming larger from outer to inner, 8-21, 2.5-4 cm long, deciduous; petals 6-12, free or basally connate; ovaries 3-locular, rarely 5-locular; styles 3-parted or apically 3-cleft. Capsules 3-locular, rarely 5-locular, 1-5 seeds per locule.

Section *Heterogena* created by J. R. Sealy, as is implied by the name, is not a natural grouping. For example, *C. granthamiana* has large and persistent bracts and sepals, 5-locular ovaries and does not belong in this subgenus but in the more primitive subgenus *Protocamellia* where *C. yunnanensis* also belongs. *C. paucipunctata* belongs to subgenus *Thea* because it is pedicellate, bracts and sepals are differentiated and the fruit has persistent bracts. *C. furfuracea* as understood by Sealy is not close to any species in his section *Heterogena*. Seven additional species have now been discovered that are similar to *C. furfuracea*, and they form a natural group which constitutes section *Furfuracea*. *C. tenii* and *C. kissii* should be assigned to section *Paracamellia*. They are very close to section *Oleifera* but the flowers are smaller, stamens fewer and shorter, petals nearly free, pedicels very short, fruit tiny, only 1 locule develops. Because this section is very confused, the species have to be placed in different sections and section *Heterogena* is no longer applicable.

7 sections, 73 species.
Type: *C. japonica* L.

SECTION IV.
Oleifera Chang, Acta Sci. Nat. Univ. Sunyatseni, monogr. ser. 1: 22. 1981.

Flores subsessiles albi 5-10 cm diam., bracteis sepalisque coriaceis indistinctis 8-11 involucratis deciduis, petalis 5-13, leviter connatis, staminibus numerosis 3-5-seriatis 1-1.5 cm longis subliberis vel leviter connatis, ovario 3-locularii, stylis 1 cm longis.

Flowers terminal, sessile, white, 5-10 cm in diameter; bracts and sepals not clearly differentiated, coriaceous, usually 8-11, involucre deciduous; petals 5-13, basally slightly connate; stamens in many series, 1-1.5 cm long, basally slightly adnate with petals; ovaries 3-locular, pubescent, rarely 5-locular; styles 1 cm long.

4 species; 3 species in China, the remaining 1 in Japan and the Ryukyu Islands.
Type: *C. oleifera* Abel

The distinguishing characteristic of section *Oleifera* is that the perules are not differentiated, deciduous after blooming; petals white, nearly free, 2.5-3.5 cm long;

stamens relatively long, more or less free; ovaries densely pubescent; styles relatively long, free. Capsules 3-5-locular, pubescent.

KEY TO SECT. *OLEIFERA*

1. Branchlets glabrous.
    2. Leaves relatively large, 5-8 cm long; bracts and sepals 10-12; flowers large, 6-7.5 cm in diameter; styles 3-5-cleft, capsules 4-7 cm in diameter, pericarp 6-8 mm thick . . . . 1. *C. gauchowensis* Chang
    2. Leaves small, 3-5.5 cm long; bracts and sepals 7-9; flowers small, 4.6 cm in diameter; styles 3-cleft, capsules 2-3 cm in diameter . . . . 2. *C. sasanqua* Thunb.
1. Branchlets pubescent.
    3. Leaves large, greater than 10 cm long; styles 3-5-cleft, petals 5 cm long . . . . 3. *C. vietnamensis* Huang ex Hu
    3. Leaves less than 10 cm long, petals 3-3.5 cm long, styles 3-parted.
        4. Leaves oblong, subglabrous, thinly coriaceous, 3-7 cm long, 2-3 cm wide; fruit 2-3 cm wide . . . . 2. *C. sasanqua* Thunb.
        4. Leaves obovate or elliptic, somewhat pubescent, thickly coriaceous, 4-9 cm long, 3-4 cm wide; fruit 3-4 cm wide . . . . 4. *C. oleifera* Abel

1. ***Camellia gauchowensis*** Chang, Bull. Sun Yatsen Univ. Nat. Sci. 1961(4): 58. 1961.

Branchlets glabrous. Leaves elliptic, 8 cm long, 4.5 cm wide, glabrous. Flowers white, 6-7.5 cm in diameter; bracts and sepals 10-12, petals 7-8, stamens free, ovaries 3-5-locular; styles 5-parted, free. Capsules pyriform, 7 cm long, 5-6 cm wide, seeds 1-4 per locule, valves 8 mm thick.

Guangdong: Gaozhou, Zhang Hongda (H. T. Chang) 5001, 5002 (holotype in SYS). Distribution: western Guangdong, wild and cultivated.

This species is differentiated from *C. oleifera* by having glabrous branchlets, more perules, larger flowers and capsules, pear shaped capsules, and a thicker pericarp.

2. ***Camellia sasanqua*** Thunb., Fl. Jap., 273. 1784.

Thea sasanqua (Thunb.) Cels, Cat. Arbres, Arbustes, 35. 1817.
Sasanqua vulgaris Nees in Siebold, Nippon 2: 13. 1832.
Sasanqua odorata Raf., Sylva Tellur., 140. 1838.

Japan: SYS 76034; Tokyo, Huang and Su (42)4; Nagasaki, May 1939, Migo H. s.n. (JSBI no. 51427); Satsuma Izumi-gun, Akune-cho, 2 January 1938, Nakajima Kazuo s.n. (JSBI no. 51429).

Distribution: Japan, occasionally cultivated in China.

This species is very close to *C. oleifera* only the leaf blades are smaller and slightly thinner, leaf apices obtuse, margins serrulate; sepals glabrous; styles shorter. This species is possibly a geographic variety of *C. oleifera*.

3. ***Camellia vietnamensis*** Huang ex Hu, Acta Phytotax. Sin. 10: 138. 1965.

Branchlets pubescent. Leaves elliptic, 5-12 cm long, 2-5 cm wide. Flowers white, 7-11 cm in diameter, bracts and sepals 9, petals 5-7, stamens slightly connate, ovaries 3-5-locular, styles 3-5-parted or basally connate. Capsules globose, 4-6 cm in diameter.

Guangxi: Liuzhou, Huang Zuojie (T. C. Huang) 2042 (holotype in PE); Luchuan, Huang Zuojie (T. C. Huang) 2033; Nanning, cultivated at the Forestry Institute, Liang Shengye (S. Y. Liang) 6403514; Nanning, Zhang Hongda (H. T. Chang) 6650, 6651; Daqing Shan, Zhang Hongda (H. T. Chang) 6688, 6689.

The leaf and flower size of this species is relatively large. Those growing at 800-950 m in the mountains have oblong leaves, to 12 cm long. Those growing at lower elevations have elliptical leaves, leaf blades thicker, wrinkled in the dry state, veins usually obscure.

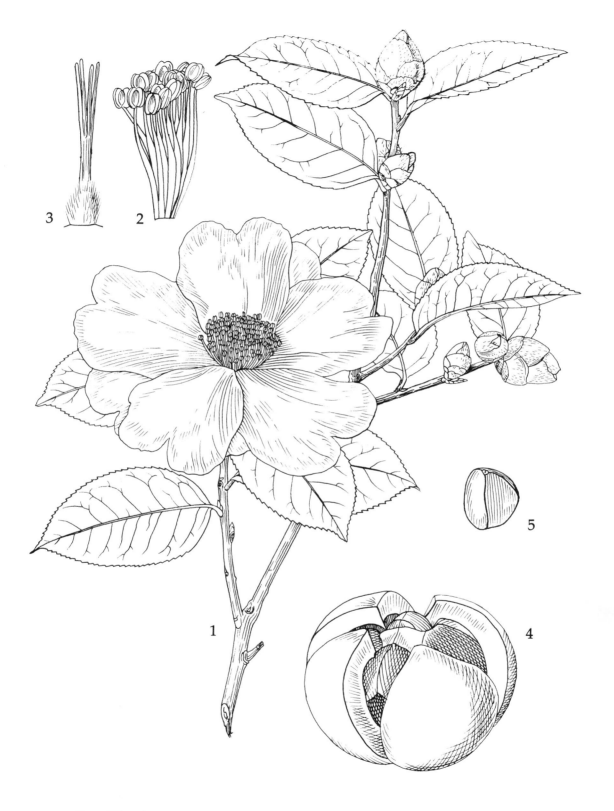

Camellia vietnamensis Huang ex Hu
1. flowering branch; 2. stamens; 3. pistil; 4. capsule; 5. seed.

4. *Camellia oleifera* Abel, Journ. China, 174, 363. 1818.
    Camellia drupifera Lour., Fl. Cochinch., 411. 1790.
    Thea podogyna Le¼v., Sert. Yunnan, 2. 1916.
    Camellia oleosa (Lour.) Wu in Engler, Bot. Jahrb. 71: 192. 1940.
    For additinal synonyms see Sealy, Rev. Gen. Camellia, 204. 1958.
    Widely cultivated in China in the provinces south of the Changjiang (Yangtse River), very variable. Leaves usually shiny, some dull in the dry state. Flowers variable in size, largest 8-9 cm in diameter, usually 5-6 cm in diameter; stamens usually free, sometimes connate into a 5-7 mm long filament tube; styles 3-parted, sometimes 4- or 5-parted. Capsules 3-locular, rarely 4-locular, but 1-locular forms also occur, usually 3.5-4.5 cm in diameter, pericarp 3-6 mm thick. In the work by Sealy (1958) the form illustrated is not typical but rather the small fruited form.

SECTION V.
Furfuracea Chang, Acta Sci. Nat. Univ. Sunyatseni, monogr. ser. 1: 25. 1981.

Flores 1-2 terminales sessiles 2-4 cm diam., perulatis, perulis (bracteis sepalisque) 8-10, deciduis, petalis 6-12 liberis vel leviter connatis, staminibus 2-3-seriatis subliberis, ovario 3-locularii, capsulis furfuraceis, seminibus 1-5 in quoque loculo.

Flowers 1-2 terminal, sessile, moderately large, 2-4 cm in diameter; 8-10 undifferentiated bracts and sepals, deciduous; petals 6-12, free or slightly connate; stamens in 2-3 series, nearly free; ovaries 3-locular, pubescent; styles 3-parted or 3-cleft. Capsules furfuraceous, 1-5 seeds per capsule.

8 species; all occurring in China, 1 species extending to Vietnam and Laos.
Type: *C. furfuracea* (Merr.) Cohen-Stuart
This section is very close to section *Oleifera*, only the perules are nearly membranous and easily broken, stamens in 2-3 series, styles usually free; pericarp spongy, furfuraceous, brown. Conversely, section *Oleifera* has coriaceous perules, stamens in 3-5 series; styles connate; pericarp woody, not furfuraceous but smooth.

KEY TO SECT. *FURFURACEA*

1. Leaves entire, lanceolate, 6-8 cm long .... 1. *C. integerrima* Chang
1. Leaf margins serrate.
    2. Petals 12; leaves oblong, veins conspicuously sunken; one seed per locule ... . 2. *C. polypetala* Chang
    2. Petals 6-9; leaves elliptic to oblong, veins usually not sunken; 3-5 seeds per locule, rarely 1 seed per locule.
        3. Leaf blades relatively large, largest often exceeding 10 cm long.
            4. Leaf bases rounded, petioles 3-5 mm long .... 3. *C. latipetiolata* Chi
            4. Leaf bases cuneate, petioles exceeding 1 cm.
                5. Flowers large 7-10 cm in diameter; fruit large, 6-10 cm in diameter, pericarp 1.5-2.5 cm thick .... 4. *C. crapnelliana* Tutch.
                5. Flowers small, 3-4 cm in diameter; fruit small, 3-4 cm in diameter.
                    6. Leaves oblong to lanceolate, pericarp 1-3 mm thick, fruit globose .... 5. *C. furfuracea* (Merr.) Cohen-Stuart
                    6. Leaves oblong to obovate-oblong, pericarp 6-7 cm thick, fruit oblate .... 6. *C. oblata* Chang ex Chang
        3. Leaves relatively small, 5-8 cm long.
            7. Leaf bases cuneate, lateral veins 5-6 pairs, oblique .... 7. *C. gaudichaudii* (Gagnep.) Sealy
            7. Leaf bases rounded, lateral veins 7-8 pairs, nearly perpendicular to midvein .... 8. *C. parafurfuracea* Liang ex Chang

1. *Camellia integerrima* Chang, Acta Sci. Nat. Univ. Sunyatseni, monogr. ser. 1: 26. 1981.

Species *C. furfuracea* proxima, a qua differt ramulis in sicco argetatis, foliis oblongo-lanceolatis minoribus intererrimis, seminibus 1-2 in quoque loculo.

Frutex circ. 2 m altus, ramulis glabris in sicco argentatis, ramis cinereo-brunneis. Folia coriacea oblongo-lanceolata, 6.5-8 cm longa 2-3 cm lata, apice acuminata basi subrotundata vel obtusa, supra nitida subtus brunnea glabra minute muriculata, nervis lateralibus 6-7-jugis utrinque conspicuis, integra, petioli 7-10 mm longi glabri. Flores non visi. Capsulae globosae vel leviter compresso-globosae 2.5-3.4 cm latae 2-2.5 cm longae, 3-loculares, 3-valvatae, valvis 2-2.5 mm crassis; semina 1-2 in quoque loculo, semiorbiculares brunnea.

Shrub 2 m tall; branchlets glabrous, silvery-white in the dry state, shiny, old branches ash-brown. Leaves coriaceous, oblong-lanceolate, 6.5-8 cm long, 2-3 cm wide, apices acuminate, bases slightly rounded or obtuse; dark green above in the dry state, shiny, glabrous; dark brown below, glabrous, somewhat muricate; lateral veins 6-7 pairs, equally conspicuous above and below; margins entire; petioles 7-10 cm long, glabrous. Flowers not seen. Capsules globose or slightly compressed, 2.5-3.4 cm wide, 2-2.5 cm high, 3-locular, 1-2 seeds per locule, 3-valvate dehiscent, valves 2-2.5 mm thick, extremely furfuraceous; columella tricornate, with narrow wings, 2-2.5 cm long; seeds semi-orbicular, brown.

Guangdong: Huiyang, Lianhua Shan, Shicheng Village, 25 August 1935, Zeng Huaide (W. T. Tsang) 25602 (holotype in SYS, isotype in A).

This species is the only one in section *Furfuracea* with entire leaves, in addition the leaf bases are subortundate and the petioles are rather long. From these characters this species can be differentiated from the other species in the section.

2. *Camellia polypetala* Chang, Acta Sci. Nat. Univ. Sunyatseni, monogr. ser. 1: 27. 1981.

Species *C. furfuracea* affinis, sed nervis et nervulis impressis, bracteis sepalisque 9, petalis 12, semine in quoque loculo solitarii distincta.

Frutex, ramulis glabris. Folia crasse coriacea anguste oblonga 10-14 cm longa 3-4.5 cm lata, apice acuminata basi late cuneata vel obtusa, supra nitida subtus nitidula glabra, nervis lateralibus 9-10-jugis ut retis supra impressis, margine serrulata, dentis 2-2.5 mm remotis, petiolis 5-7 mm longis glabris. Flores terminales vel axillares 1-2-flori circ. 3.5 cm diam. albi sessiles; bracteis et sepalis 9 exterioribus late ovoideis 1-1.5 cm longis extus pubescentibus, basi subliberis, intimis petaloideis; petalis 12, extimis coriaceis margine scariosis, 15-17 mm longis, interioribus obovatis circ. 20 mm longis 10-15 mm latis glabris vel apicem versus pubescentibus, basi connatis; staminibus 14 mm longis basi connatis glabris; ovariis sericeis, stylis 3 liberis 8-10 mm longis pubescentibus. Capsulae globosae 20-25 mm in diametro extus furfuraceae 3-valvatae, valvis 1.5 mm crassis, columella triangulata 20 mm longa, semina solitaria in quoque loculo 15-17 mm longa brunnea.

Shrub 4 m tall, branchlets glabrous. Leaves thickly coriaceous, narrowly oblong, 10-14 cm long, 3-4.5 cm wide, apices acuminate, bases broadly cuneate or slightly rounded; dark green above in the dry state, shiny; light brown below, glabrous, shiny; lateral veins 9-10 pairs, reticulate veins sunken above; margins serrulate, teeth separation 2-2.5 mm; petioles 5-7 mm long, glabrous. Flowers 1-2 terminal or axillary, 3.5 cm in diameter, sessile; bracts and sepals 9, 4 outermost perules broadly ovoid, 1-1.5 cm long, exterior pubescent, bases nearly free; innermost perules becoming petal-like; petals 12, 2 outermost petals coriaceous in the middle, pubescent, margins thinly membranous, 15-17 mm long; inner 10 petals broadly obovate, 20 mm long, 10-15 mm wide, glabrous, or pubescent near the apices, basally connate for 3-4 mm; stamens 14 mm long, basally connate for 3-4 mm, glabrous; anthers minute, yellow, ovaries sericeous; styles 3-parted, completely free, 8-10 mm long, pubescent. Capsules globose, 20-25 mm in diameter, outer surface furfuraceous, 3-locular, 1 seed per locule,

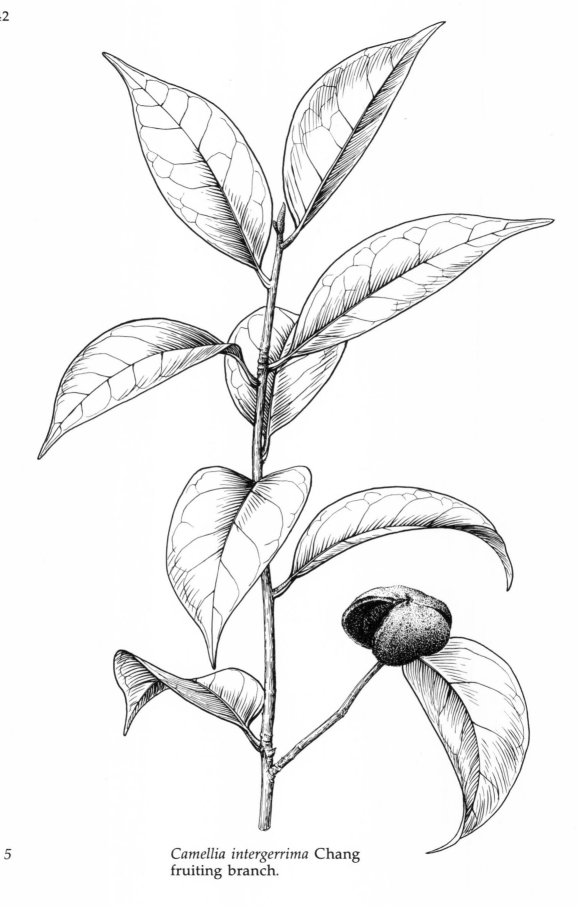

5 *Camellia intergerrima* Chang fruiting branch.

pericarp 1.5 mm thick; columella triangular, 20 mm long; seeds hemispherical, 15-17 mm long, brown.

Guangdong: Fengkai Xian, Yufu, Huanggang Shan, Zhenshuping, by a stream, shrub 4 m tall, flowers white, 11 December 1963, Chen Shaoqing (S. C. Chun) 18404 (holotype in SCBI); Fengkai, Heishiding, Zhu Zhisong (C. S. Chu) 50057.

3. *Camellia latipetiolata* Chi, Sunyatsenia 7: 18. 1948.

This species is very close to *C. furfuracea* only, the leaves are larger and thinner, bases rounded and petioles shorter.

Guangdong: Xinyi, Huang Zhi (C. Wang) 37764 (holotype in SCBI).

Distribution: western Guangdong.

4. *Camellia crapnelliana* Tutch., Journ. Linn. Soc. London 37: 63. 1904.

Camellia gigantocarpa Hu & Huang in Hu, Acta Phytotax. Sin. 10: 133. 1965.
Camellia octopetala Hu, Acta Phytotax. Sin. 10: 135. 1965.
Camellia latilimba Hu in Anonymous, Icon. Comophyt. Sin. 2: 854. 1972, *nom. illegit.*

Bark red, smooth, branchlets glabrous. Leaves elliptic, 8-12 cm long, 4-5 cm wide, glabrous. Flowers large, 8-10 cm in diameter, bracts and sepals 8, petals 6-8; stamens slightly connate at base, 1.5-1.8 cm long; ovaries pubescent, 3-locular; styles 3-parted, free. Capsules globose, 7-10 cm in diameter; pericarp slightly spongy, 1.5-2.5 cm thick, exterior furfuraceous; 3-5 seeds per locule.

Hong Kong: Tutcher 967 (type fragment); Hong Kong Herbarium, Y. S. Lau 1328; Hong Kong Dalong Farm, Li Fusheng (F. S. Lai) s.n. (SYS no. 149083 in fruit).

Guangxi: Bobai, Huang Zuojie (T. C. Huang) 2026 (holotype of *C. gigantocarpa* in PE); Nanning, GXFI Arboretum, Zhang Hongda (H. T. Chang) 6646, 6648; same loc., Liang Shengye (S. Y. Liang) 6403507; Pubei, Zhou Wenzhi (W. E. Chou) 69970.

Zhejiang: Longquan, HZBG 7002 (holotype of *C. octopetala* in PE), 5010.

Fujian: Dehua, Zhong Buqin (P. C. Tsoong) 72; Wu Kemin (K. M. Wu) 60291; Huang Shumei (S. M. Huang) 190673.

This species is very close to *C. furfuracea*. The latter often has oblong leaves; small flowers, 3-4 cm in diameter; pericarp thin, 1.5 mm thick. Since 1904, other than the type (Tutcher 967), no one has described this species and no one knew the character of the fruit. Thus it was a partially understood species. The author saw a specimen of this species with fruit in the Hong Kong government Herbarium in June 1979 and determined this specimen to be in section *Furfuracea*. Neither *C. gigantocarpa* or *C. octopetala* described by Hu, can be differentiated from *C. crapnelliana*.

5. *Camellia furfuracea* (Merr.) Cohen-Stuart, Bull. Jard. Bot. Buitenzorg, ser. 3, 1: 240. 1919.

Thea furfuracea Merr., Philip. Journ. Sci. Bot. 13: 149. 1918.
Theopsis furfuracea (Merr.) Nakai, Journ. Jap. Bot. 16: 706. 1940.
Thea bolovensis Gagnep., Not. Syst. 10: 124. 1942.
Camellia furfuracea var. *lutea* Hu, Acta Phytotax. Sin. 8: 266. 1963.
Camellia pubisepala Fang, Acta Bot. Yunnanica 2(3): 337. 1980.

Leaves relatively variable, usually obovate, thinly coriaceous, some plants have leaves oblong or narrowly lanceolate, width variable. Size of the fruit and thickness of the valves is variable which has caused confusion in this species. However, the flower shape is more fixed. Flowers 3-3.5 cm in diameter; perules 7-9, thinly membranous, easily broken; petals 7-8, ca. 1.5 cm long, nearly free or slightly connate; stamens 1.2-1.5 cm long, basally connate; ovaries 3-locular, pubescent; styles 3-parted, free, ca. 1.5 cm long, pubescent. Capsules globose or oblate, 2.5-3.5 cm in diameter; pericarp reddish-brown, 2-3 mm thick, furfuraceous.

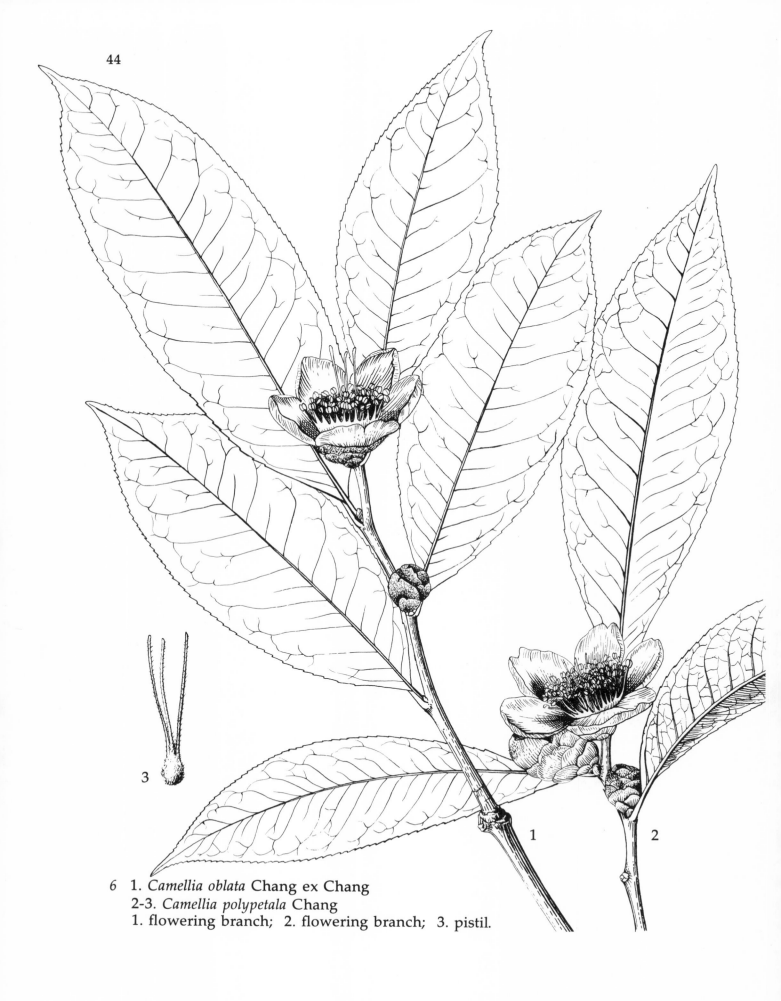

6 1. *Camellia oblata* Chang ex Chang
2-3. *Camellia polypetala* Chang
1. flowering branch; 2. flowering branch; 3. pistil.

Hainan: Diaoluo Shan, Exped. Eastern-Hainan Acad. Sin. 904 (holotype of *C. furfuracea* var. *lutea* in SCBI).

Guangdong: Wengyuan, Liu Xinqi (S. K. Lau) 24521; Yingde, Liang Xiangyue (H. Y. Liang) 61386, 60500. The three specimens above have straight lanceolate leaves.

Guangxi: Rongxian, Tiantang Shan, Chen Shaoqing 9619 (leaves lanceolate, bases round); Gengcheng, Shiwan Dashan, in evergreen forest, elev. 310 m, 16 November 1979, Qin Dehai and Li Chunting 76658 (holotype of *C. pubisepala* in GXMI).

Hunan: Yingzhang, Mang Shan, Chen Shaoqing (S. C. Chun) 2568, 3555; same loc., Liang Baohan 83916; same loc., Liu Linhan (H. L. Liu) 1201, 1258. The leaf veins of the above specimens are sunken.

Jiangxi: Anyuan, HLG 2274.

Vietnam: Dahuangmao Shan, Zeng Huaide (W. T. Tsang) 27328, 30708.

Distribution: China, Vietnam and Laos.

6. ***Camellia oblata*** Chang ex Chang, *sp. nov.;* Chang, Acta Sci. Nat. Univ. Sunyatseni, monogr. ser. 1: 30. 1981, *nom. invalid.*

Species affinis *C. furfuracea,* sed foliis punctatis, capsulis oblatis, pericarpio 6-7 mm crasso differt.

Frutex, ramulis glabris. Folia oblonga vel oblongo-oblanceolata 12-15 cm longa 4-5.5 cm lata, apice abrupte acuta basi late cuneata, supra nitida, subtus bruneo-viridia glabra minute atro-punctata, nervis lateralibus circ. 7-jugis supra conspicuis subtus elevatis, margine serrulata vel inferiore subintegra, petioli 7-12 mm longi glabri. Flores albi, bracteis et sepalis 10 scariosis deciduis; petalis 9; staminibus 9 mm longis subliberis; ovariis pilosis, stylis 4 liberis 1-1.2 cm longis. Capsula sessilis compresse globosa 3.2 cm lata 2 cm longa, pericarpio 6-7 mm crasso extus furfuraceo, sepalis persistentibus suborbicularibus 7-8 mm longis extus pubescentibus.

Shrub 3 m tall, branchlets glabrous. Leaves coriaceous, oblong to oblong-oblanceolate, 12-15 cm long, 4-5.5 cm wide, apices abruptly acute, bases broadly cuneate; light green above in the dry state, shiny, glabrous; light brown below, glabrous, minutely atropunctate; lateral veins 7 pairs, conspicuous above, protruding below; margins serrulate or lower half nearly entire; petioles 7-12 mm long, glabrous. Flowers white; bracts and sepals 10, scarious, early deciduous; petals 9; stamens 9 mm long, nearly free; styles 1-1.2 cm long, 4-parted, free, pubescent. Capsules terminal, sessile, oblate, 3.2 cm wide, 2 cm tall, not yet ripe, pericarp 6-7 mm thick, exterior very furfuraceous; sepal remnants 2-3, nearly orbicular, 7-8 mm long, exterior pubescent.

Guangxi: Shiwan Dashan, elev. 1000 m, 14 August 1933, Zuo Jinglie (C. L. Tso) 23646 (*holotypus* in SCBI); GXFI 5357.

7. ***Camellia gaudichaudii*** (Gagnep.) Sealy, Kew Bull. 1949(2): 217. 1949.

Camellia hongkongensis Seem., Trans. Linn. Soc. London 22: 342. 1859, *quoad* Gaudichaud 271.

Thea gaudichaudii Gagnep., Not. Syst. 10: 127. 1942.

Branchlets glabrous. Leaves elliptic, stiffly coriaceous, apices obtuse, back atropunctate. Sepals deciduous, ovaries pubescent; styles 3-parted, free. Capsules globose, furfuraceous.

Hainan: Wenchang Xian, Feng Qin (H. Fung) 20336.

Guangxi: Dongxing, Xu Peilai (P. L. Liao) 7502; Dongxing, Zhong Yecong (Y. C. Chung) 625.

Jiangxi: Xinwu, Shangping Commune, Yang Xiangxue (C. X. Yang) 12458.

Distribution: Southern China, also seen in Vietnam.

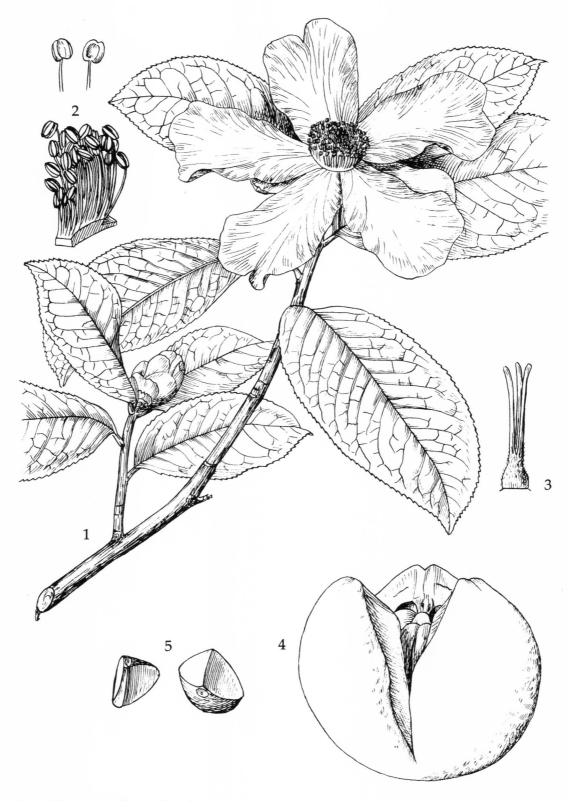

7 *Camellia crapnelliana* Tutch.
   1. flowering branch; 2. stamens; 3. pistil; 4. capsule; 5. seeds.

Camellia parafurfuracea Liang ex Chang 8
fruiting branch.

8. *Camellia parafurfuracea* Liang ex Chang, Acta Sci. Nat. Univ. Sunyatseni, monogr. ser. 1: 31. 1981.

A *C. furfuracea* et affinitate differt foliis minoribus nervis lateralibus subperpendicularibus, floribus minoribus, capsulis tenuiter valvatis, semine in quoque loculo solitarii.

Frutex, ramulis glabris. Folia elliptica coriacea 6-8.5 cm longa 2.8-3.7 cm lata, apice acuminata basi rotundata vel obtusa, supra viridia nitida subtus glabra, nervis lateralibus 7-8-jugis, margine calloso-serrulata, petioli 5 mm longi glabri. Flores 1-3 axillares vel terminales albi sessiles; bracteis et sepalis 9-10 sericeis, 3-10 mm longis deciduis; petalis 7-8 obovatis vel ovatis 1.2-1.4 cm longis 8-11 mm latis, apice rotundatis vel obtusis basi connatis staminibus; 8-11 mm longis, basi 2-3 mm ad petalam adnatis, filamentis liberis glabris; ovariis pilosis, stylis 3 liberis circ. 1 cm longis pubescentibus. Capsula globosa 2.5-3 cm diam. 3-locularis, pericarpio 2.5 mm crasso furfuraceo, columella 1.7 cm longa, semina solitaria in quoque loculo globosa vel semiglobosa 1.7 cm longa brunnea.

Shrub, branchlets glabrous. Leaves coriaceous, elliptic, 6-8.5 cm long, 2.8-3.7 cm wide, apices acuminate, bases rounded or obtuse; dark green above in the dry state, shiny; light brown below, glabrous; lateral veins 7-8 pairs, margins corniculate-serrulate; petioles ca. 5 mm long, glabrous. Flowers 1-3 axillary or terminal, white, sessile; bracts and sepals 9-10, exterior sericeous, interior glabrous, deciduous after anthesis; lowest 2 perules suborbicular, 3-4 mm long; remaining 7-8 perules broadly ovoid, 7-10 mm long; petals 7-8 obovate or ovate, 1.2-1.4 cm long, 8-11 mm wide, apices rounded or obtuse, bases connate for 2 mm; stamens 8-11 mm long, glabrous, basally adnate with petals for 2-3 mm, free filaments glabrous; ovaries pilose; styles 3-parted, 1 cm long, pilose. Capsules subglobose, furfuraceous, 2.5-3 cm in diameter, 3-locular, 1 seed per locule, 3-valvate dehiscent, pericarp 2.5 mm thick, pedicels very short, columella 1.7 cm long; seeds subglobose or hemispherical, 1.7 cm long, brown.

Guangxi: Napo Xian, 26 October 1972, Liang Shengye (S. Y. Liang) 721029 (holotype in SYS).

Guangdong: Huiyang, Changan Village, Wei Zhaofen (C. F. Wai) 12158; Haifen, Gaotan, Geobot. Lab. SCBI 7603.

SECTION VI.
Paracamellia Sealy, Rev. Gen. Camellia, 192. 1958.

Flowers small, 2-3 cm in diameter, rarely larger, sessile; bracts and sepals undifferentiated, 6-10, early deciduous; petals 5-8, nearly free; stamens short, in 1-2 series, slightly connate; ovaries 3-locular, pubescent; styles 3-parted, 2-7 mm long. Capsules small, 1-2 cm in diameter, usually 1-locular, rarely 2-3-locular, 1 seed per locule.

16 species; 14 in China, the other 2 species occurring in Indochina, India and the Ryukyu Islands.

Type: *C. kissii* Wall.

This section is close to section *Oleifera*, but the flowers are smaller, petals nearly free, stamens in 1-2 series, pedicels shorter, capsules usually 1-locular and 1 seed per locule. Also the growth habit is a small shrub.

KEY TO SECT. *PARACAMELLIA*

1. Leaves 5-11 cm long, usually elliptic, rarely narrowly lanceolate.
    2. Leaves elliptic or oblong, 2-5 cm wide.
        3. Flowers large, 5 cm in diameter; leaves 7-10 cm long.
            4. Leaves oblong, sharply serrate; capsules globose, 2-2.5 cm in diameter
            .... 1. *C. grijsii* Hance

4. Leaves elliptic, serrate; capsules ovate, 1.5 cm wide ... 2. *C. confusa* (Cohen-Stuart)
3. Flowers relatively small, 2-3.5 cm in diameter, sometimes slightly larger; leaves 5-7 cm long, oblong or elliptic.
    5. Filaments nearly free, stamens 1 cm long, bracts and sepals 9-10, petals 7-8 .... 3a. *C. kissii* Wall. var. *kissii*
    5. Filaments 5-6 mm long, connate for 4/5 of length; bracts and sepals 8 .... 4. *C. lutescens* Dyer
2. Leaves narrowly lanceolate, 1-2 cm wide .... 5. *C. fluviatilis* Hand.-Mazz.
1. Leaves 2-6 cm long.
    6. Leaves elliptic, ovate or obovate, 2 times as long as broad.
        7. Leaves elliptic or oblong.
            8. Branchlets pubescent, petals 1-3 cm long.
                9. Flowers single, white, petals 1-1.5 cm long, stamens and pistals normally developed.
                    10. Bracts and sepals 6-7, pedicels with 2-3 scars left by the bracts and sepals; leaves coriaceous, narrow elliptic, apices obtuse; petals 5 .... 6. *C. brevistyla* (Hay.) Cohen-Stuart
                    10. Bracts and sepals 10, pedicels with 5-6 ringlike marks; leaves broadly elliptic, apices obtuse .... 7. *C. obtusifolia* Chang
                9. Flowers double, red, stamens and pistils abnormal, petals 3 cm long .... 8. *C. maliflora* Lindl.
            8. Branchlets glabrous, petals 1-2.5 cm long.
                11. Flowers white (excluding cultivars).
                    12. Leaves oblong, lateral veins not sunken in the dry state, margins subentire .... 9. *C. miyagii* (Koidz.) Mak. & Nem.
                    12. Leaves broadly elliptic, margins sharply serrate, veins sunken .... 10. *C. shensiensis* Chang ex Chang
                11. Flowers red or pink, 3-4 cm in diameter.
                    13. Leaves oblong, slightly shiny; petals oblong .... 11. *C. brevissima* Chang & Liang
                    13. Leaves elliptic, extremely shiny; petals obovate .... 12. *C. puniceiflora* Chang
        7. Leaves obovate or ovate.
            14. Leaves ovate, acute, 4.5 cm long, margins sparsely serrate; bracts and sepals 9, petals 6 .... 13. *C. tenii* Sealy
            14. Leaves obovate, apices rounded or obtuse, margins densely serrate; bracts and sepals 5-7, petals 5-7 .... 14. *C. microphylla* (Merr.) Chien
    6. Leaves narrowly oblong or oblanceolate, 3-4 times as long as wide.
        15. Flowers small, 2-3 cm in diameter.
            16. Bracts and sepals 9-10, leaf apices acute.
                17. Petals 5; leaf apices acute, lower surfaces pubescent .... 15. *C. phaeoclada* Chang
                17. Petals 7; leaf apices abruptly acute, lower surfaces glabrous .... 3a. *C. kissii* Wall. var. *kissii*
            16. Bracts and sepals 7, leaf apices obtuse .... 6. *C. brevistyla* (Hay.) Cohen-Stuart
        15. Flowers large, 5-6 cm in diameter.
            18. Sepals and petals pubescent, leaf apices acute .... 3b. *C. kissii* var. *megalantha* Chang
            18. Sepals and petals glabrous, leaf apices obtuse or obtusely pointed ... . 16. *C. weiningensis* Y. K. Li ex Chang

1. *Camellia grijsii* Hance, Journ. Bot. 17: 9. 1879.
   Thea grijsii (Hance) O. Kuntze, Rev. Gen. Pl., 65. 1891.
   Camellia yuhsienensis Hu, Acta Phytotax. Sin. 10: 139. 1965.
   Leaves oblong, 6-9 cm long. Flowers 4-5 cm in diameter, bracts and sepals 9-10, petals 5-6, stamens 7-8 mm long, styles 3-4 mm long. Capsules globose, 2 cm in diameter, 1-3-locular, pericarp thin.
   Fujian: Sha Xian, Lin Ying 02; Jianning, Lin Laiguan 830; Zhong Xinxuan 3479; Wang Dashun 136.
   Hunan: Yongshun, Wen Xuankai 800703; You Xian, Liu Jinpu s.n. (SYS no. 140928).
   Jiangxi: Lichuan, Hengdian, Wang Mingjin 2059.
   Guangxi: Longshen, Zhong Jixin 91074.
   Yunnan: Kunming, Zhang Hongda (H. T. Chang) 5772 (cultivated).
   Distribution: Fujian, Jiangxi, Guangxi, Hunan, Hubei.
   Zhong Jixin 91047 has long ovate leaves, bases nearly rounded, the rest of the characters are the usual ones for this species.

2. *Camellia confusa* (Craib) Cohen-Stuart, Meded. Proefst. Thee 40: 71. 1916.
   Thea confusa Craib, Kew Bull. 1914(1): 5. 1914.
   C. oleifera Abel var. *confusa* (Craib) Sealy, Rev. Gen. Camellia, 209. 1958.
   This species is distinguished from *C. oleifera* by the bracts and sepals being scarious, pedicels shorter; capsules minute, elliptic, less than 2 cm long, 1-locular, valves thin, only 1 seed.
   Yunnan: Mengla, Wang Qiwu (C. W. Wang) 80177, 80223; Jinghong, Wang Qiwu (C. W. Wang) 75687; Lincang, Wang Qiwu (C. W. Wang) 76677; Liushun, Wang Qiwu (C. W. Wang) 81099; Sino-Soviet Expedition 5458, 7692.
   Guangxi: Shiwan Dashan, Qin Renchang (R. C. Ching) 8171.
   Distribution: Yunnan, Guangxi, Thailand, Laos, Burma.

3. *Camellia kissii* Wall., Asiat. Res. 8: 429. 1820.
   Camellia symplocifolia Griff., Itin. Notes 40, no. 652. 1848.
   Camellia mastersia Griff., Notul. 4: 559. 1854.
   Thea sasanqua var. *kissii* (Wall.) Pierre, Fl. Forest. Cochinchine 2: sub. pl. 115-116. 1887.
   Thea iniquicarpa (Clarke) Kochs in Engler, Bot. Jahrb. 27: 594. 1900.
   Thea bachmaensis Gagnep., Not. Syst. 10: 124. 1942.
   Thea brachystemon Gagnep., Not. Syst. 10: 124. 1942.
   For additional synonyms see Sealy, Rev. Gen. Camellia, 197. 1958.

3a. *Camellia kissii* Wall. var. *kissii*
   Leaves very variable, from elliptic to oblong or oblanceolate, veins sunken in the dry state. Flowers small, petals nearly free, stamens basally connate, ovaries pubescent; styles 3-parted or 3-cleft, 5-7 mm long. Capsules obovate, 1-3-locular.
   Guangxi: Qin Renchang (R. C. Ching) 8062, 8120; Shiwan Dashan, Zhang Zhaoqian 12411; Ningming, Zhang Zhaoqian 12395; Shiwan Dashan, Zeng Huaide (W. T. Tsang) 22027, 24506, 24609.
   Guangdong: Xinan, Xu Deming 254; Landao Island, McClure 1192; Luofu Shan, Jiang Ying (Y. Tsiang) 1711; Yingde, Dazhen, Zeng Huaide (W. T. Tsang) and Huang Jingzhou 2445; Yingde, Wentang Shan, Chen Huan 7122; Conghua, Deng Liang 8430, 8431; Ying De, Huashui Shan, Zengpei, Feng Qin (H. Fung) 10957; Luofu Shan, Gao Xiping 52484; Longmen, Li Xuegen 200385.
   Hainan: Ganen, Liu Xinqi 27777; Baoting, Liu Xinqi 28085.
   Vietnam: M. Poilane 17472.
   Distribution: Yunnan, Guangxi, Guangdong, Hainan, Nepal, India, Sikkim and Indochina.

3b. ***Camellia kissii*** var. ***megalantha*** Chang, Acta Sci. Nat. Univ. Sunyatseni, monogr. ser. 1: 35. 1981.

A typo differt floribus multo majoribus, petalis 3 cm longis 2 cm latis.

Flowers relatively large, 6 cm in diameter; petals obcordate, 3 cm long, 2 cm wide, other characters the same as the species.

Guangxi: Zhaoping, Nanyong Village, 11 December 1965, Liang Shengye (S. Y. Liang) 6505258 (holotype in SYS); Guilin, Liangfeng, Deng Zhinong 13220.

4. ***Camellia lutescens*** Dyer in Hook. f., Fl. Brit. India 1: 293. 1874.

Thea lutescens (Dyer) Pierre, Fl. For. Cochinchine 2: sub pl. 119. 1887.

Branchlets glabrous. Leaves oblong, 5-9 cm long, 2-3 cm wide, glabrous. Bracts and sepals 8, semi-persistent; petals 6, filaments highly connate; ovaries pubescent; styles 3, free, extremely short.

Distribution: Assam in eastern India.

Within this section this species has the most connate filaments. The sepals are deciduous in young fruit which is the common condition seen in this section.

5. ***Camellia fluviatilis*** Hand.-Mazz., Anz. Akad. Wiss. Math. Nat. Wien 59: 57. 1922.

Thea fluviatilis (Hand.-Mazz,) Merr., Lingnan Sci. Journ. 7: 316. 1931.
Camellia stenophylla Kobuski, Brittonia 4: 115. 1941.
Camellia kissii var. *stenophylla* (Kobuski) Sealy, Rev. Gen. Camellia, 201. 1958.

Branchlets puberulent. Leaves narrowly lanceolate. Flowers small, petals nearly free, filaments basally slightly connate, ovaries pubescent; styles short, 3-parted.

Guangdong: Yingde, R. Mell 44 (photograph of holotype); no loc., G. Ford s.n. (Hong Kong Botanical Garden no. 4841).

Hainan: Liu Xinqi (S. K. Lau) 28372; Zhang Haidao 2379.

Guangxi: Shansi, Zhang Zhaoqian 13191; Zhaoping, Li Zhongdi 60231; Zhaoping, Liang Shengye (S. Y. Liang) 6401.

Distribution: China, Burma and India.

6. ***Camellia brevistyla*** (Hay.) Cohen-Stuart, Meded. Proefst. Thee 40: 67. 1916.

Thea brevistyla Hay., Fl. Mont. Formos., 63. 1908.
Thea tenuiflora Hay., Journ. Coll. Sci. Tokyo 30: 46. 1911.
Thea gnaphalocarpa Hay., Icon. Pl. Formos. 3: 44. 1913.
Camellia tenuiflora (Hay.) Cohen-Stuart, Meded. Proefst. Thee 40: 68. 1916.
Camellia gnaphalocarpa (Hay.) Cohen-Stuart, Meded. Proefst. Thee 40: 68. 1916.
Theopsis lungyaiensis Hu, Acta Phytotax. Sin. 10: 141. 1965.

Branchlets pubescent. Leaves oblong or elliptic, petioles short. Bracts and sepals 7-8, petals 5, filaments basally slightly connate, ovaries pubescent; styles 3-4-parted, 1.5-3 mm long. Capsules globose, 1 cm in diameter.

Taiwan: Mt. Minami-Soten, 1919, S. Sasaki s.n. (Taiwan For. Res. Inst. Herb. no. 16876); Taibei, Suzuki-Tokio 18965, 7829, 8678, 19413, 14250.

Fujian: Shaowu, Zhou Hechang 6541.

Guangdong: Ruyuan, Liu Xinqi (S. K. Lau) 29197; Ruyuan, Huang Zhi (C. Wang) 42446; Ruyuan, Mou Ruhuai 40069, 40071, 40072, 40074.

Anhui: Qimen, Deng Moubin 5216; Qin Renchang (R. C. Ching) 3108.

Jiangxi: Wugong Shan, PE Jiangxi Team 599, 959; Longnan, Jiangxi Univ. 12167.

Guangxi: Liang Shengye (S. Y. Liang) s.n. (SYS no. 137624).

Distribution: Fujian, Taiwan, Guangdong, Guangxi, Anhui, Jiangxi.

The bark is reddish-brown on the old branches of Zhou 6541 from Fujian. Qin 3108 from Anhui was determined by Sealy (Rev. Gen. Camellia, 215. 1958) to be *C. microphylla,* but it belongs to this species.

9 1-2. *Camellia fluviatilis* Hand.-Mazz. 3-4. *Camellia kissii* Wall.
1. fruiting branch; 2. pistil; 3. fruiting branch; 4. pistil.

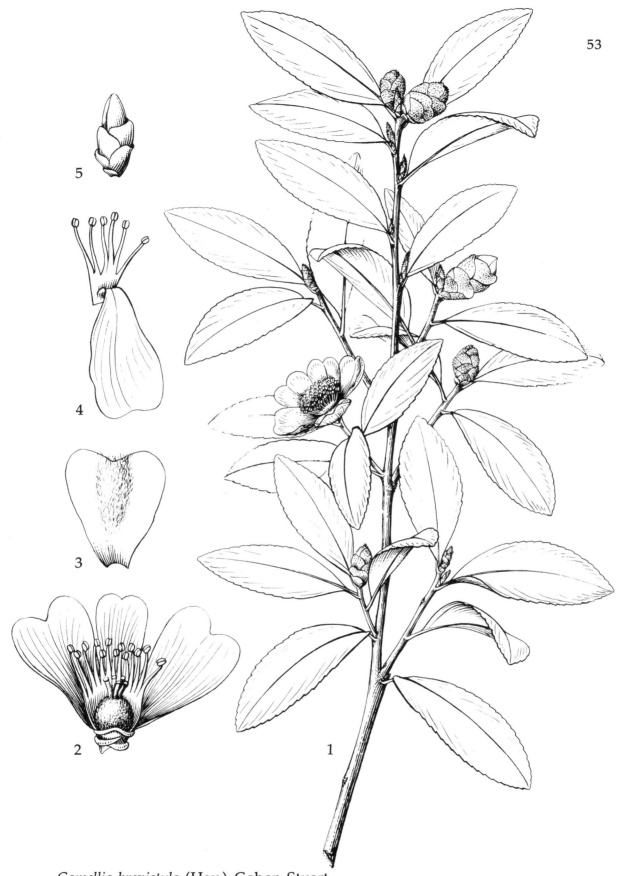

Camellia brevistyla (Hay.) Cohen-Stuart
1. flowering branch; 2. flower; 3. petal; 4. petal and stamens; 5. flower bud.

7. **Camellia obtusifolia** Chang, Acta Sci. Nat. Univ. Sunyatseni, monogr. ser. 1: 38. 1981.

Species C. brevistylae affinis, sed foliis late ellipticis obtusis, floribus brevibus, petalis 1-1.2 cm longis apice rotundatis, stylis longioribus 7-8 mm longis differt.

Frutex vel arbor parva, ramulis hirsutis. Folia late elliptica 3.5-5 cm longa 2.5-3 cm lata, apice obtusa vel subrotundata basi late cuneata vel subrotundata, supra nitida ad costam puberula subtus glabra, nervi laterales untrinsecus 6, margine serrulata, petioli 3-4 mm longi pubescentes. Flores bini terminales albi sessiles, bracteis et sepalis 10, semilunatis vel obovatis, 2-8 mm longis, glabris vel ciliatis; petalis 7(5) obovatis, 1-1.2 cm longis, 7-9 mm latis, apice rotundatis glabris, basi subliberis; staminibus 2-seriatis, exterioribus basi ⅓ connatis, 1 cm longis glabris; ovariis hirsutis, stylis 3 liberis 7-8 mm longis glabris.Capsula globosa 1.5-2 cm in diametro 1-3 locularis, 3-valvata dehiscens, valvis tenuibus circ. 1 mm crassis; semina solitaria in quoque loculo.

Shrub or small tree 4 m tall; branchlets hirsute, old branches glabrous. Leaves broadly elliptic, 3.5-5 cm long, 2.5-3 cm wide, apices obtuse or nearly rounded, bases broadly cuneate or slightly rounded; dark green above, midveins puberulent; yellowish-brown below, glabrous; lateral veins 6 pairs, margins serrulate; petioles 3-4 mm long, pubescent. Flowers usually 2, terminal, white, sessile; bracts and sepals 10, crescent shaped to obovate, 2-8 mm long, glabrous, margins with long cilia; petals 7(5), obovate, 1-1.2 cm long, 7-9 mm wide, apices rounded, glabrous, basally nearly free; stamens 1 cm long, in 2 series, outer whorl connate for the basal ⅓, glabrous, inner whorl relatively fewer stamens, completely free; ovaries hirsute; styles 3-parted, 7-8 mm long, glabrous.Capsules globose, 1.5-2 cm in diameter, 3-locular or 1-locular, 1 seed per locule, 3-valvate dehiscent; pericarp thin, less than 1 mm thick; bracts and sepals not persistent.

Fujian: no loc., Tang Ruiyong 252 (holotype in SYS).

Zhejiang: Longquan, Zhang Shaoyao 6939.

Jiangxi: Lichuan, Fengping, October 1957, Wang Mingjin 2247; Guangchang, Jiangxifeng Shan, Hu Qiming 5322.

Guangdong: Ruyuan, Tisha Shan, Huang Zhi (C. Wang) 42446; same loc., Gao Ximing 53396.

8. **Camellia maliflora** Lindl., Bot. Reg. 13: sub. pl. 1078. 1827.
Camellia sasanqua Sims, Bot. Mag. 46: sub. pl. 2080. 1819, *auct. non* Thunb.
Sasanqua malliflora Raf., Sylva Tellur., 140. 1838.
Thea maliflora (Lindl.) Seem., Trans. Linn. Soc. London 22: 346. 1859.
Theopsis maliflora (Lindl.) Nakai, Journ. Jap. Bot. 16: 706. 1940.

Shrub, branchlets pubescent. Leaves membranous, elliptic, 3.5-5 cm long, 1.8-3 cm wide, apices acute, bases broadly cuneate or slightly rounded, midvein pubescent above, lower surfaces pubescent, margins serrulate; petioles 3-5 mm long, pubescent. Flowers red, double; pedicels 4-5 mm long; bracts and sepals 9-10, hemispherical to orbicular, 1.5-8 mm long, deciduous after anthesis, outer surfaces puberulent; petals 3 cm long, petals of outer whorl nearly orbicular, 2.3 cm long; petals of inner whorl formed from the stamens, relatively short and narrow; stamens abnormal, 1 cm long; ovaries deformed, glabrous; styles 3 or more parted, free or connate.

Cultivated at Kew, Hort. Bot. Reg. Kew, 31 January 1935 and 13 January 1936.

This species was reported to have been introduced to England from China in 1819, but the species is not now found in China. From the abnormal ovaries and stamens and its sterility, the species is obviously a hybrid. J. R. Sealy in his monograph put this species in section *Theopsis*. Because of the undifferentiated perules which are deciduous after anthesis, this species should be in section *Paracamellia*. The author has seen the specimens at Kew and discussed the question of the position of this species with Sealy. Sealy agrees with the present position.

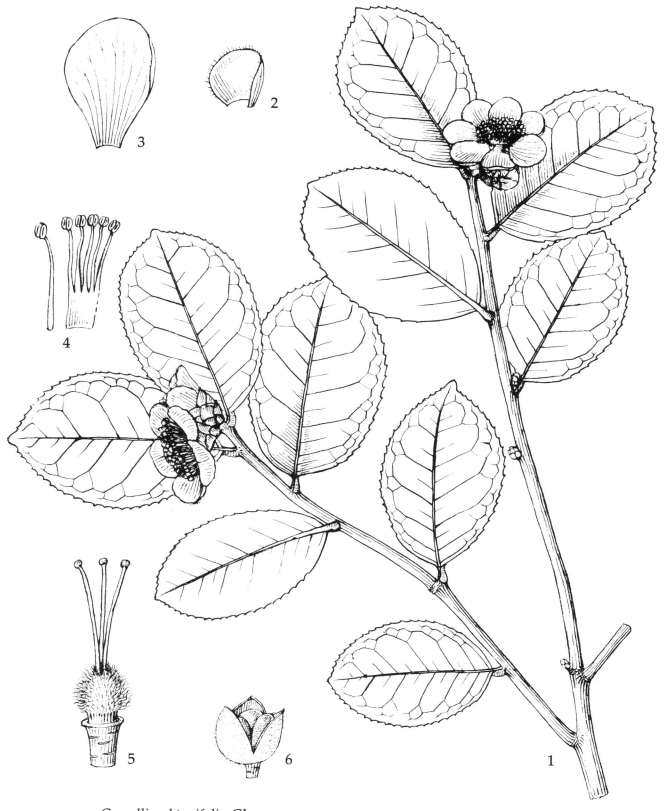

Camellia obtusifolia Chang
1. flowering branch; 2. perule; 3. petal; 4. stamens; 5. pistil; 6. capsule.

9. *Camellia miyagii* (Koidz.) Mak. & Nem., Fl. Jap., ed., 2, 738. 1931.
Thea miyagii Koidz., Bot. Mag. Tokyo 32: 252. 1918.
This species is very close to *C. brevistyla*, only the branchlets are glabrous; leaves slightly larger, margins not serrate, petals pubescent; styles 5-6 mm long, upper half deeply 3-cleft. Cultivated cultivars are double flowered, styles 3 mm long.
Guangdong: Guangzhou, cultivated, Chen Shaoqing (S. C. Chun) 7028.
Japan: Ryukyu Islands, 17 December 1937, Suzuki Toioi s.n. (JSBI no. 51422).

10. *Camellia shensiensis* Chang ex Chang, *sp. nov.;* Chang, Acta Sci. Nat. Univ. Sunyatseni, monogr. ser. 1: 39. 1981, *nom. invalid.*

A *C. obtusifolia* differt foliis acuriter densiusque serratis, apice abrupte acutis basi cuneatis glabris punctatis, sepalis longioribus circ. 6-10 mm longis, stylis brevioribus.

Frutex circ. 1.5 m altus, ramulis pubescentibus. Folia coriacea late elliptica 3.5-5 cm longa 2-3 cm lata, apice abrupte acuta basi late cuneata, supra nitida subtus glabra atropunctata, nervi laterales circ. 5-jugi ut retis nervorum utrinque conspicui, margine acriter serrulata, petioli 3-5 mm longi. Flores singulari vel bini terminales et axillares albi sessiles; bracteis et sepalis 7-8 circ. 6-10 mm longis sparse pubescentibus; petalis 6-7 glabris 1.5-2 cm longis, apice bilobatis, basi connatis; staminibus 6-8 mm longis subliberis; ovariis pilosis 3-locularibus, stylis 3-4 liberis 3 mm longis. Capsula ovoidea 1.5-1.8 cm longa.

Shrub 1.5 m tall, branchlets pubescent. Leaves coriaceous, broadly elliptic, 3.5-5 cm long, 2-3 cm wide, apices abruptly acute, bases broadly cuneate, shiny above, glabrous and glandular punctate below, lateral veins 5 pairs, lateral and reticulate veins conspicuous from both surfaces, margins sharply serrulate, petioles 3-5 mm long. Flowers 1-2 terminal or axillary, sessile, white; bracts and sepals 7-8, 6-10 mm long, sparsely pubescent; petals 6-7, glabrous, 1.5-2 cm long, apically 2-cleft, basally connate; stamens 6-8 mm long, nearly free; ovaries pubescent, 3-locular; styles 3 mm long, 3-4-parted, free. Capsules ovoid, 1.5-1.8 cm long.

Sichuan: Beipei, Jinyun Shan, Yao Zhongwu 3410 (*holotypus* in PE); Tongjian Xian, Wang Zongjiao 70.
Shaanxi: Qiao Yinglin 1140.
Hubei: Xingshan suburbs, Wuhan Botanical Garden 1860.
Yunnan: Cultivated at Heilongtan, introduced from Sichuan, Kunming Work Station Xinjingsan 50453; same loc., Zhang Hongda (H. T. Chang) 5122.

This species has the northernmost distribution of *Camellia* in China. The southern slopes of the Qingling Range has subtropical pockets in which this species occurs in Sichuan and western Hubei.

11. *Camellia brevissima* Chang & Liang in Liang, Acta Phytotax. Sin. 20: 116. 1982.
Species affinis *C. sasanquae*, sed differt floribus minoribus, puniceis vel rubellis, petalis anguste obovato-ellipticis.

Frutex sempervirens, 1.5-5 m altus, cortice cinereo-brunneo vel flavo-brunneo; ramuli hornotini teretes, flavo-brunnei, annotini cinereo-brunnei glabri. Folia coriacea, elliptica vel ovato-oblonga, 3-5.5 cm longa, 1.5-2.5 cm lata, apice acuminata, basi cuneata, utrinque costa nervisque exceptis glabra, nervis lateralibus et venulis reticulatis utrinsecus inconspicuiis, margine in ⅔ vel ½ superiore serrulata; petioli 3-6 mm longi pubescentes. Flores in alabastro punicei, sub anthesi rubelli, sessiles vel subsessiles; bracteis et sepalis 7-8, rotundis viridibus deciduis; petalis 5, anguste obovato-ellipticis 15-20 mm longis, 5-10 mm latis, apice emarginatis; staminibus numerosis, filamentis rubellis, basi ad petalam adnatis; antheris flavescentibus; ovariis subglobosis 3-locularibus, cinereo-albidis pubescentibus, stylis 3 mm longis, apice 3-fidis. Capsula subglobosa, 1.5-2.5 cm in diametro, pericarpio tenui; semina 1-2 in quoque loculo, nigrescentibrunnea.

Shrub 1.5-5 m tall, bark greyish-brown or yellowish-brown; branchlets terrete, yellowish-brown; older branches greyish-brown, glabrous. Leaves alternate, coriaceous, elliptic or ovate-elliptic, 3-5.5 cm long, 1.5-2.5 cm wide, apices acuminate, bases cuneate, both surfaces equally glabrous, only the midvein pubescent, lateral and reticulating veins equally obscure on both surfaces, upper ⅓ to ½ of margins serrulate; petioles pubescent, 3-6 mm long. Flowers usually solitary, terminal or axillary, bud ovate, before opening reddish-purple, after opening becoming red or pink, sessile or subsessile; bracts and sepals 7-8, rotundate, green, deciduous; petals usually 5, narrowly obovate-elliptic, 15-20 mm long, 5-10 mm wide, apices emarginate; filaments light-red, basally adnate with petals; stamens many, arranged in several series, anthers yellow; ovaries subglobose, 3-locular, greyish-white pubescent; pedicels 3 mm long, stigma shallowly 3-parted. Capsules small, suborbicular, 1.5-2.5 cm in diameter, pericarp thin; seeds 1-2 per locule, rarely three, blackish-brown. Flowering and fruiting period October to December.

Guangxi: He Xian, Gupo Shan, 27 November 1965, Liang Shengye (S. Y. Liang) 6505245 (holotype in GXFI; isotypes in SYS, PE, GXMI); same loc., Liang Shengye (S. Y. Liang) 7801119.

12. *Camellia puniceiflora* Chang, Acta Sci. Nat. Univ. Sunyatseni, monogr. ser. 1: 40. 1981.

A C. *brevistyla* differt foliis ellipticis lucidissimis, floribus puniceis, stylis longioribus; a C. *obtusifolia* ramulis glabris, foliis lucidissimis, floribus majoribus puniceis, petalis longioribus circ. 3 cm longis, pericarpio crassiore 3-4 mm crasso recedit.

Frutex circ. 2 m altus, ramulis glabris. Folia coriacea elliptica 4-5 cm longa 2-2.5 cm lata, apice obtusa vel subacuta basi late cuneata, supra lucidissima subtus glabra, nervis lateralibus utrinsecus 5-6, margine serrulata, petiolis 3-4 mm longis pubescentibus. Flores terminales et axillares sessiles puniceis circ. 6 cm in diametro; bracteis et sepalis 7-8, exterioribus circ. 1 cm longis ovatis sparse pubescentibus; petalis 5-6 obovatis 3 cm longis basi leviter connatis; staminibus 1-1.3 cm longis liberis glabris; ovariis pilosis, stylis 3 liberis glabris 5-7 mm longis. Capsula globosa 1-locularis, pericarpio 3-4 mm crasso.

Shrub 2 m tall, branchlets glabrous. Leaves coriaceous, elliptic, 4-5 cm long, 2-2.5 cm wide, apices obtuse or slightly acute, bases broadly cuneate, extremely shiny above, glabrous below, lateral veins 5-6 pairs, margins serrate; petioles 3-4 mm long, pubescent. Flowers terminal or axillary, sessile, pink, 6 cm in diameter; bracts and sepals 7-8, to 1 cm long, ovoid, exterior slightly pubescent; petals 5-6, obovate, 3 cm long, slightly connate; stamens 1-1.3 cm long, free, glabrous; ovaries pubescent; styles 3-parted, glabrous, 5-7 mm long. Capsules globose, 1-locular, pericarp 3-4 mm thick.

Zhejiang: Longquan, Jinqi, in a valley by a stream, 10 November 1959, flowers pink, Zhang Shaoyao 7058, 7150 (holotype in HZBG); Tianmu Shan, He Xianyu 25168; HZBG no. 1821.

13. *Camellia tenii* Sealy, Kew Bull. 1949(2): 221. 1949.
Theopsis euonymifolia Hu, Acta Phytotax. Sin. 10: 140. 1965.

Branchlets pubescent. Leaves ovate-orbicular, 1.5-4.5 cm long, 1.2-2.4 cm wide, veins pubescent below. Bracts and sepals 9, deciduous; petals 6, stamens basally connate; styles 3-parted, 4.6 mm long. Capsules globose, 1-1.5 cm wide, 3-locular, 1 seed per locule, valves 1-5 mm thick.

Yunnan: Liu Shenne (S. N. Liu) 22476, 17348, 22481, 22652; Dayao, Wood Oil Group 65-318 (in fruit); Jingdong, Li Minggang 1506 (*Theopsis euonymifolia* holotype in PE).

This species is very close to C. *brevistyla*, only the leaves are ovate-orbicular, thinner, and the styles are slightly longer.

12 *Camellia shensiensis* Chang ex Chang
1. flowering branch; 2. petals and stamens; 3. petal; 4. pistil; 5. bud.

1-5. *Camellia tenii* Sealy 6. *Camellia puniceiflora* Chang 7. *Camellia microphylla* (Merr.) Chien
1. flowering branch; 2. flower bud; 3-4. stamens and pistil; 5. stamens;
6. flowering branch; 7. flowering branch.

14. *Camellia microphylla* (Merr.) Chien, Contrib. Biol. Lab. Sci. Soc. China Bot. 12: 100. 1939.

   Thea microphylla Merr., Journ. Arn. Arb. 8: 9. 1927.
   Theopsis microphylla (Merr.) Nakai, Journ. Jap. Bot. 16: 706. 1940.
   Branchlets pubescent. Leaves narrow and small, 2-3 cm long, 1.3 cm wide, apices rounded or obtuse. Bracts and sepals 6-7; petals 5-7, 1 cm long; basal half of stamens connate; styles 3-parted, 2-3 mm long.
   Guizhou: Yingjaing, Jian Zhuopo 30040.
   Anhui: Qimen, Deng Moubin 5143.
   Hunan: Chengbu, Tan Peixiang 63862.
   Jiangxi: Lichuan, Wang Mingjin 2326.

15. *Camellia phaeoclada* Chang, Acta Sci. Nat. Univ. Sunyatseni, monogr. ser. 1: 41. 1981.

   Species *C. brevistylae* affinis, a qua differt foliis anguste oblongis 4-5.5 cm longis 1.3-1.8 cm latis, apice acutis, staminibus subliberis, stylis longioribus, connatis brevius 3-fidis.
   Frutex circ. 2.5 m altus, ramulis brunneo-pubescentibus. Folia coriacea anguste oblonga 4-5.5 cm longa 1.3-1.8 cm lata, apice acuta basi cuneata, supra nitida basi bruneo-viridia primo pubescentia nox glabrescentia, nervis lateralibus circ. utrinsecus 5 utrinque inconspicuis, margine serrulata, petioli 3-4 mm longi puberuli. Flores terminales sessiles albi 2.5-3 cm diam.; bracteis et sepalis 10 deciduis obovatis 5-9 mm longis glabris vel ciliatis subacutis; petalis 5 circ. 1.5 cm longis; staminibus 3-seriatis subliberis 8-10 mm longis; ovariis pilosis, stylis 8-10 mm longis glabris apice 3-fidis. Capsula globosa 1.5 mm in diametro 1-locularis 3-valvate dehiscens, valvis tenuibus, pedicellis brevissimis; semina globosa 1.2 cm diam. atrobrunnea.
   Shrub 2.5 m tall; branchlets with brown pilose pubescence, old branches brown. Leaves coriaceous, narrowly oblong, 4-5.5 cm long, 1.3-1.8 cm wide, apices acute, bases cuneate, shiny above; yellowish-brown below, at first villous, later glabrous; lateral veins 5 pairs, equally obscure from both surfaces; margins serrulate; petioles 3-4 mm ong, pubescent. Flowers terminal, sessile, white, 2.5-3 cm in diameter; bracts and sepals 10, deciduous, obovate, 5-9 mm long, exterior glabrous, margins ciliate, apices slightly pointed; petals 5, 1.5 cm long; stamens in 3 series, nearly free, 8-10 mm long; ovaries pubescent; styles glabrous, 8-10 mm long, apices shallowly 3-cleft. Capsules globose, 1.5 cm in diameter, 1-locular, 1 seed per locule; seeds globose, 1.2 cm in diameter, dark brown; fruit shell thin, 3-valvate dehiscent; pedicels relatively short.
   Sichuan: Huidong Xian City to Yangfa Brigade, Dai Tianlun 11823.
   Yunnan: Shuangjiang, First District, between Taiping and Mengwen, elev. 1200 m, shrub, flowers white, fruit red, 26 September 1957, Xin Jingsan (G. S. Sin) 1245; Shuangbai, Yin Wenqing 861 (holotype in SCBI); Weishan Xian, Yang Zenghong 6967.
   Distinguished from *C. brevistyla* by narrow oblong leaves, narrower and longer and with apices acute.

16. *Camellia weiningensis* Y. K. Li ex Chang, *sp. nov.;* Y. K. Li in Chang, Acta Sci. Nat. Univ. Sunyatseni, monogr. ser. 1: 42. 1981, *nom. invalid.*

   A *C. brevistyla* differt foliis rigidis, floribus majoribus 4-6 mm longis.
   Frutex, ramulis glabris. Folia coriacea oblonga 4-6 cm longa 1.6-2 cm lata, apice obtusa vel subacuta basi late cuneata, utrinque in sicco nitida glabra, nervis lateralibus utrinsecus 6-7 impressis, margine serrulata, petioli 4-6 mm longi. Flores singulari vel bini subterminales sessiles albi vel rubelli; bracteis et sepalis 6-7 coriaceis glabris vel ciliatis exterioribus 1.5 cm longis; petalis 6-8 inaequalibus obovatis 1.5-3 cm longis glabris basi 7 mm connatis; staminibus 2-3-seriatis 6-9 mm longis inferioribus connatis glabris; ovariis sericeis, stylis 6-8 mm longis glabris apice 3-fidis. Capsula compresse tricocca 2.5-3 cm lata 1.5-2 cm alta, 3-locularis, pericarpio circ. 3 mm crasso, semina

Camellia phaeoclada Chang fruiting branch.

15 *Camellia weiningensis* Y. K. Li
1. flowering branch; 2. petal and stamens; 3. stamens; 4. pistil; 5 capsule; 6. seed.

singula in quoque loculo.

Shrub, branchlets glabrous. Leaves coriaceous, oblong, 4-6 cm long, 1.6-2 cm wide, apices obtuse or slightly acute, bases broadly cuneate; both surfaces shiny in the dry state, glabrous; lateral veins 6-7 pairs, sunken in the dry state; margins densely serrulate, petioles 4-6 mm long. Flowers 1-2 terminal, sessile, pink or white; bracts and sepals 6-7, coriaceous, glabrous, or ciliate, to 1.5 cm long; petals 6-8, unequal, obovate, 1.5-3 cm long, glabrous, basally connate for 7 mm; stamens in 2-3 series, 6-9 mm long, lower half connate, glabrous; ovaries sericeous; styles 6-8 mm long, glabrous, apically 3-cleft. Capsules tricoccus-oblate, 2.5-3 cm wide, 1.5-2 cm high, 3-locular, pericarp 3 mm thick.

Guizhou: Weining, Wang Daozhi s.n. (SYS nos. 141864, 142129 (*holotypus* in SYS, *isotypus* in Agric. Coll. Guizhou)).

SECTION VII.
Pseudocamellia Sealy, Rev. Gen. Camellia, 147. 1958.

Flowers terminal or axillary, white, nearly sessile; bracts and sepals not distinctly differentiated, membranous, 9-10, semipersistent; petals 6-9, basally connate; stamens in 3-4 series, filaments often connate; ovaries 3-locular, 1 seed per locule, pericarp smooth.

5 species in China's southwestern provinces.

Type: *Camellia szechuanensis* Chi

KEY TO SECT. *PSEUDOCAMELLIA*

1. Ovaries pubescent . . . . Ser. I. *Trichocarpae* Chang
  2. Leaves oblong or elliptic, bases cuneate.
    3. Lateral veins 8-10 pairs, protruding above . . . . 1. *C. szechuanensis* Chi
    3. Lateral veins 5-7 pairs, sunken above . . . . 2. *C. chungkingensis* Chang
  2. Leaves long-ovate, bases rounded . . . . 3. *C. trichocarpa* Chang
1. Ovaries glabrous . . . . Ser. II. *Gymnocarpae* Chang
  4. Leaf bases cuneate, blades oblong, branchlets glabrous . . . .
    . . . . 4. *C. ilicifolia* Y. K. Li ex Chang
  4. Leave bases rounded, blades long-ovate, branchlets pubescent . . . .
    . . . . 5. *C. henryana* Cohen-Stuart

SER. I.
Trichocarpae Chang, Acta Sci. Nat. Univ. Sunyatseni, monogr. ser. 1: 44. 1981.

Flores majores 4-6 cm diam., petala 7-9, ovarium tomentosum.
Flowers relatively large, 4-6 cm in diameter, petals 7-9, ovaries pubescent.
3 species in Sichuan and Yunnan.
Type: *Camellia szechuanensis* Chi

1. ***Camellia szechuanensis*** Chi, Sunyatsenia 7: 17. 1948.

Leaves with ca. 11 pairs of lateral veins. Flowers 3 cm in diameter; perules nearly membranous, easily breaking; ovaries and styles pubescent. Capsules small and with several persistent bracts.

Sichuan: Emei Shan, Yang Guanghui 51904, 51906, 50991, 53538, 53629, 53650, 53670, 53695, 53880, 53953, 55758, 57194, 56020, 56231; same loc., Zhang Hongda (H. T. Chang) 5641, 5663, 5669.

2. *Camellia chungkingensis* Chang, Acta Sci. Nat. Univ. Sunyatseni, monogr. ser. 1: 44. 1981.

A *C. szechuanensi* differt foliis ovato-oblongis brevioribus 6-9 cm longis, crassius coriaceis, obtuse acutis vel abrupte acutiusculis, nervis lateralibus paucioribus 5-7-jugis, perulis paucioribus.

Frutex vel arbor parva, ramulis glabris. Folia coriacea ovato-oblonga vel elliptica 6-9 cm longa 2.5-4.5 cm lata, apice subacuta vel obtusa interdum abrupte acuta, basi late cuneata vel subrotundata, supra nitida subtus brunneo-viridia glabra, nervis lateralibus 5-7-jugis in sicco ut retis nervorum impressis, margine serrulata, petioli 7-10 mm longi. Flores axillares et terminales albi sessiles; bracteis et sepalis 8 extus pubescentibus, margine tenuibus scariosis; petalis 7 obovatis 1.5-2.5 cm longis basi connatis extus plus minusve pubescentibus; staminibus 1-1.3 cm longis basi 3 mm ad petalam adnatis, tubo filamentorum 2-3 mm longo; ovariis pilosis, stylis 3 liberis dense cinereo-pilosis 1.6-1.8 cm longis.

Small tree or shrub; branchlets glabrous, older branches greyish-brown in the dry state. Leaves thickly coriaceous, ovate-oblong or elliptic, 6-9 cm long, 2.5-4.5 cm wide, apices slightly acute or obtuse, or abruptly acute, bases broadly cuneate or nearly rounded; dark green above in the dry state, shiny; yellowish-green or brown below; lateral veins 5-7 pairs, sunken above, slightly protruding below; reticulating veins sunken above, obscure below; margins serrulate; petioles 7-10 mm long, deeply grooved above. Flowers axillary or terminal, sessile; bracts and sepals 8, pubescent in the middle of the outer surface, inner surface glabrous, margins thin, membranous, early deciduous; petals ca. 7, obovate, 1.5-2.5 cm long, bases cuneate, outer surface somewhat pubescent; stamens 1-1.3 cm long, basal 3 mm adnate with petals; filament tube 2-3 mm long; ovaries pubescent, styles 3-parted, completely free, densely greyish-white pilose, 1.6-1.8 cm long.

Sichuan: Zhongqing, Beipei, November 1937, Qu Zhongxian 4181 (holotype in PE); same loc., Yao Zhongming 5277; Jinyun Shan Re-examination Team 081.

The leaf blades of this species are ovate-oblong, thickly coriaceous, basally slightly rounded, apices slightly obtuse, lateral veins 5-7 pairs. Only 8 perules, styles densely tomentose.

Du Tahua 48 from Jinyun Shan of Beipei has leaves oblong, not shiny in the dry state, margins with uneven blunt serrations; terminal young fruit, surface without tubercles but with 4 shallow grooves. This collection could belong to this species.

3. *Camellia trichocarpa* Chang, Acta Sci. Nat. Univ. Sunyatseni, monogr. ser. 1: 45. 1981.

C. henryana Sealy, Rev. Gen. Camellia, 161. 1958, *quoad* Henry 10908, 10908A, 10908B, *auct. non* Cohen-Stuart.

A *C. henryana* differt foliis coriaceis, floribus minoribus petalis 1.5-2.5 cm longis, ovario dense villoso.

Frutex vel arbor parva, ramulis pubescentibus. Folia coriacea ovata-oblonga 5-7.5 cm longa 2.5-3.5 cm lata, apice acuminata basi subcordata vel rotundata, supra nitidula subtus ad costam puberula, nervis lateralibus 5-7-jugis, margine serrata, petioli 4-5 mm longi pubescentes. Flores terminales albi sessiles; bracteis et sepalis 9-10 suborbicularibus scariosis pubescentibus exterioribus 8-9 mm longis; petalis 7-8 obovatis 1.5-2.5 cm longis basi paulo connatis; staminibus 1-1.4 cm longis liberis; ovariis pilosis, stylis 3-4 liberis 1.4 cm longis.

Small tree 2-3 m tall, branchlets pubescent. Leaves coriaceous, long ovate, 5-7.5 cm long, 2.5-3.5 cm wide, apices acuminate, bases subcordate or orbicular, slightly shiny above, somewhat pubescent along the midveins below, lateral veins 5-7, margins sharply serrate; petioles 4-5 mm long, pubescent. Flowers terminal, white, sessile; bracts and sepals 9-10, nearly orbicular, membranous in the dry state, pubescent, to 8-9 mm long; petals 7-8, obovate, 1.5-2.5 cm long, bases slightly connate; stamens free,

Camellia chungkingensis Chang
1. flowering branch; 2. petals and stamens; 3. pistil. *16*

1-1.4 cm long; ovaries pubescent; styles 3-4-parted, free, ca. 1.4 cm long.

Yunnan: Nanjian, Xia Lifang 79 (holotype in KUN); Nujiang river valley, Yu Dejun (T. T. Yü) 22997; Jingdong, Li Minggang 2511.

When Cohen-Stuart published *C. henryana* in 1916 he combined specimens with glabrous and pubescent ovaries. Sealy designated Henry 10908 as the lectotype, and this specimen has glabrous ovaries. We believe that the specimens with glabrous and pubescent ovaries belong to different species.

SER. II.
Gymnocarpae Chang, Acta Sci. Nat. Univ. Sunyatseni, monogr. ser. 1: 46. 1981.

Flores minores 2.5-4 cm diam., petala circ. 6, ovarium glabrum.
Flowers relatively small, 2.5-4 cm in diameter, petals ca. 6, ovaries glabrous.
2 species distributed in Guizhou, Yunnan and Sichuan.
Type: *Camellia henryana* Cohen-Stuart

4. *Camellia ilicifolia* Y. K. Li ex Chang, *sp. nov.;* Y. K. Li in Chang, Acta Sci. Nat. Sunyatseni, monogr. ser. 1: 46. 1981, *nom. invalid.*

Ramulis glabris. Folia oblonga 6-9.5 cm longa 2.5-3.6 cm lata, apice acuminata basi late cuneata, nervis lateralibus 6-8-jugis, margine acute serrata, petiolis 6-8 mm longis. Flores albi; bracteis et sepalis 8 subscariosis 1 cm longa; petalis 6, 1.7-2 cm longis basi connatis; ovariis glabris 3-locularis, stylis 3 liberis 1.2 cm longis coronans.

Shrub or small tree, branchlets glabrous. Leaves coriaceous, oblong, 6-9.5 cm long, 2.5-3.6 cm wide, apices acuminate, bases broadly cuneate, lateral veins 6-8 pairs, margins sharply serrate, petioles 6-8 mm long. Flowers white, sessile; bracts and sepals 8, somewhat membranous, to 1 cm long; petals 8, 1.7-2 cm long, bases connate; stamens 1.1-1.5 cm long, outer filament whorl connate, glabrous; ovaries globose, 3-locular; styles 3-parted, 1.2 cm long, free.

Guizhou; Chishui, Jinsha Gou, Nan Forest Center, Yezhuping, Li Yongkang 74357 (*holotypus* in SYS, *isotypus* in Guizhou Agric. Coll.); Zhijin Forest Center, Ghishi Team 98.

5. *Camellia henryana* Cohen-Stuart, Meded. Proefst. Thee 40: 132. 1916.
Thea henryana (Cohen-Stuart) Rehder, Journ. Arn. Arb. 5: 338. 1924.

Branchlets pubescent. Leaves ovate-elliptic, 7-11 cm long, 3-4.5 cm wide, bases rounded. Ovaries 3-locular, glabrous; styles 3-parted, free. Capsules oblate, with sepals early deciduous.

Sichuan: Huili, Yu Dejun (T. T. Yü) 1611.
Yunnan: Fugong, Cai Xitao (H. T. Tsai) 54874; Menghai, Wang Qiwu (C. W. Wang) 77304; Yu Dejun (T. T. Yü) 17762, 18258; Jingdong, Yu Shuogui 5109.
Distribution: Yunnan, Sichuan.

When Cohen-Stuart published this species, he mixed up specimens with glabrous and non-glabrous ovaries. Sealy mentioned this problem and used a glabrous specimen, Henry 10908 as the lectotype. Based on a large number of specimens, we believe that the difference between glabrous and non-glabrous ovaries is not a minor distinction. This difference is correlated with leaf thickness, the non-glabrous ovary plants having membranous leaves and those with glabrous ovaries having coriaceous leaves.

SECTION VIII.
Tuberculata Chang, Acta Sci. Nat. Univ. Sunyatseni, monogr. ser. 1: 47. 1981.

Floribus terminalibus solitariis albis subsessilibus; bracteis sepalisque 9-10 scariosis indistinctis deciduis vel sub deciduis; petalis 6-7 angusto-oblongis basi connatis; staminibus 3-seriatis filamentis connatis; ovario 3-5-locularii sulcato, stylis 3-5 liberis; capsulis tuberculatis, semine 1 quoque loculo puberulo.

Flowers solitary, terminal, white, subsessile; bracts and sepals not clearly differentiated, membranous, 9-10, deciduous or subdeciduous after flowering; petals 6-7, nearly oblong, basally connate, stamens in 3 series, filaments connate into a tube; ovaries 3-5-locular, grooved; styles 3-5-parted, free. Capsules tuberculate, 1 seed per locule, seed surfaces pubescent.

6 species, endemic to China.

Type: *C. tuberculata* Chien

KEY TO SECT. *TUBERCULATA*

1. Leaves elliptic, oblong, obovate or oblanceolate, bases cuneate.
   2. Leaves oblong, atropunctate below; ovaries 3-5-locular, styles 3-5-parted..... .... 1. *C. tuberculata* Chien
   2. Leaves obovate, obovate-oblong or oblanceolate, not atropunctate; ovaries 3-locular; styles 3-parted, glabrous.
      3. Leaves obovate or obovate-oblong, apices acute.
         4. Leaves obovate, to 5.5 cm wide, coriaceous ..... 2. *C. anlungensis* Chang
         4. Leaves oblanceolate, 3.5 cm wide, thinly coriaceous .... 3. *C. obovatifolia* Chang
      3. Leaves oblong, apices acuminate......... 4. *C. rhytidocarpa* Chang & Liang
1. Leaves ovate or long-ovate, bases rounded.
   5. Leaves large, 8 cm long, lateral veins 10 pairs; sepals glabrous.............. .... 5. *C. litchii* Chang
   5. Leaves small, 5 cm long, lateral veins 6-8 pairs; sepals tomentose .... 6. *C. parvimuricata* Chang

1. ***Camellia tuberculata*** Chien, Contrib. Biol. Lab. Sci. Soc. China Bot. 12: 94. 1939.
Thea sp. Wils. in Sargent, Pl. Wils. 3: 451. 1917.

Sichuan: Beipei, Qu Guiling (K. L. Chu) 6621; Jinyun Shan, Yang Xianjin (Y. C. Yang) 3280 (lectotype here designated in PE); Leshan, Wang Shujia 213; Fengjie, Shiziba, Fang Mingyuan 24752; Jinyun Shan, Pei Jian 75151; Nanchuan, Wang Fazuan 10901.

Fang 24752 has relatively small leaf blades, 5-7.5 cm long, 2-2.5 cm wide; styles 4-parted, free and pubescent. The typical species has 5-parted free styles, and the back of the leaves are atropunctate.

2. ***Camellia anlungensis*** Chang, Acta Sci. Nat. Univ. Sunyatseni, monogr. ser. 1: 48. 1981.

A *C. tuberculata* foliis obovatis epunctatis, capsulis majoribus 3-3.5 cm diam., 3-locularibus, pericarpio crassiori 2-4 mm crasso differt.

Frutex vel arbor parva, ramulis glabris. Folia coriacea obovata 7-14 cm longa 3.5-5.5 cm lata, apice abrupte acuta basi late cuneata, supra viridia opaca subtus flavo-virentia epunctata, nervis lateralibus utrinsecus 6-9, margine serrulata, petioli 5-7 mm longi. Flores non visi. Capsula globosa 3-3.5 cm in diametro, subsessilis, 3-locularis 3-valvata dehiscens, valvis 2-4 mm crassis extus crispis et tuberculatis pubescentibus; semina solitaria in quoque loculo semiglobosa extus tomentella.

Shrub or small tree, branchlets and terminal buds glabrous. Leaves coriaceous,

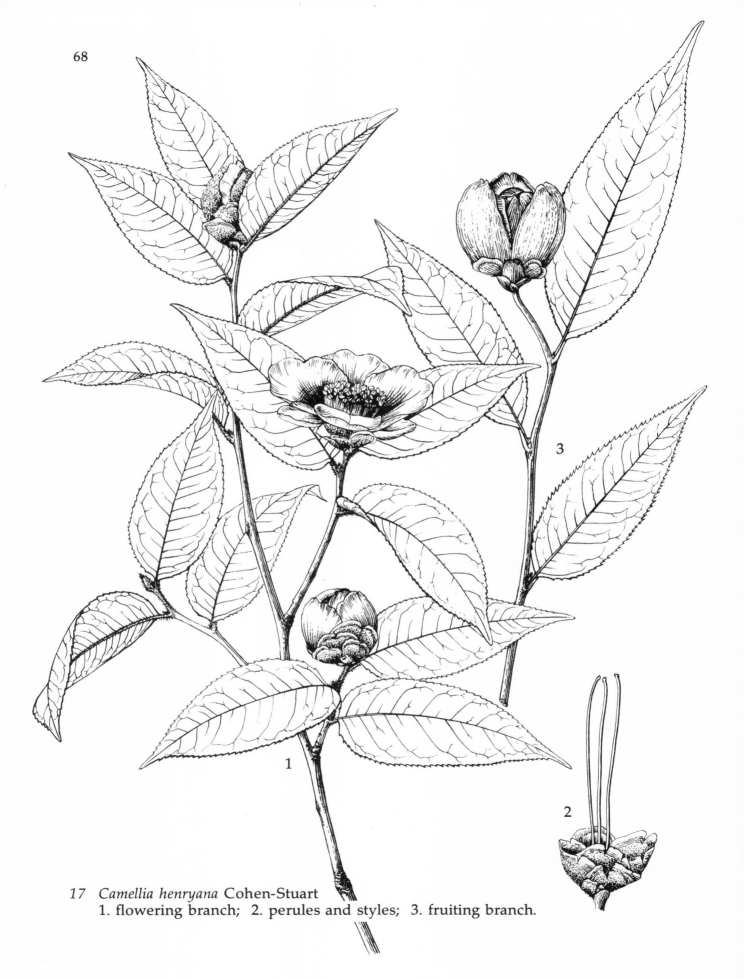

17 *Camellia henryana* Cohen-Stuart
1. flowering branch; 2. perules and styles; 3. fruiting branch.

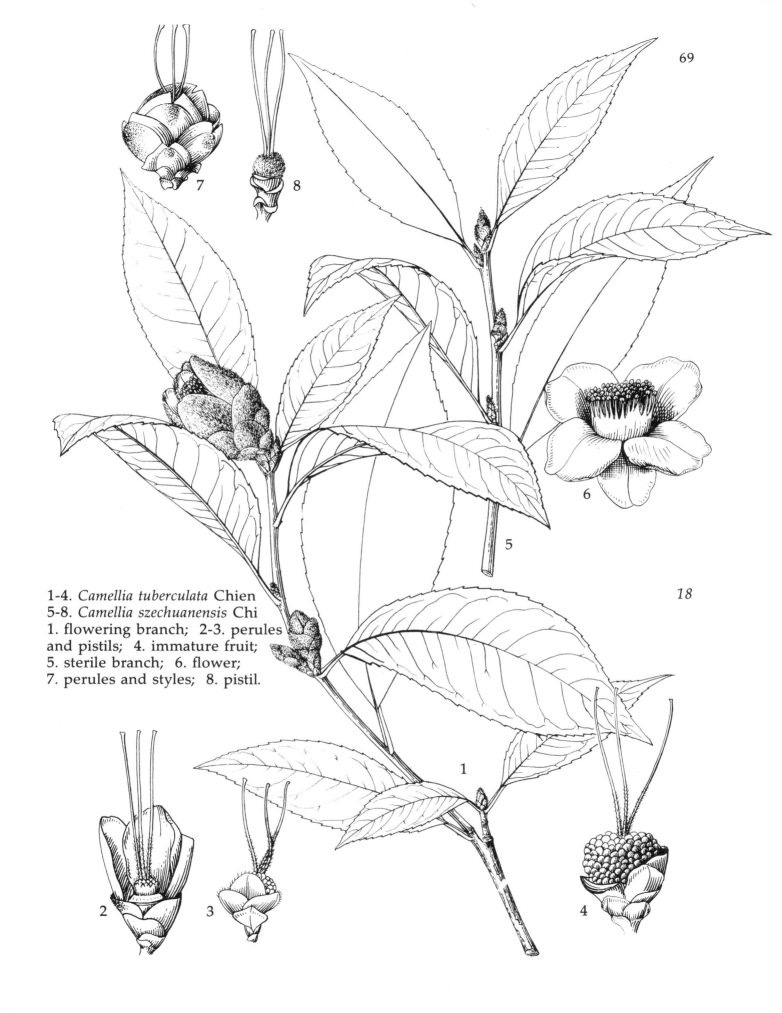

1-4. *Camellia tuberculata* Chien
5-8. *Camellia szechuanensis* Chi
1. flowering branch; 2-3. perules and pistils; 4. immature fruit; 5. sterile branch; 6. flower; 7. perules and styles; 8. pistil.

obovate, 7-14 cm long, 3.5-5.5 cm wide, apices acute, bases broadly cuneate; light green above, shiny; atropunctate below, lateral veins 6-9 pairs, margines sharply serrulate, petioles 5-7 mm long. Capsules nearly globose, sessile, 3-3.5 cm in diameter, 3-locular, 1 seed per locule, pericarp rugose or tuberculate, 3-valvate dehiscent, valves 2-4 mm thick, pubescent; seeds hemispherical, surfaces tomentose.

Guizhou: Anlong, Bot. Inst. Guizhou Team, Zhang Zhisong and Zhang Yongtian 5952; Ceheng, Luxiong Lumber Camp, Cao Ziyu 1227, 1259; Wangmu, Guizhou Team 1757 (holotype in PE).

3. *Camellia obovatifolia* Chang, Acta Sci. Nat. Univ. Sunyatseni, monogr. ser. 1: 48. 1981.

A *C. tuberculata* foliis obovato-lanceolatis, ovariis subglabris stylis 3 glabris differt; a *C. anlungensi* foliis angustioribus, ovariis et stylis glabris recidit.

Arbor parva, ramulis glabris. Folia tenuiter coriacea obovato-lanceolata vel oblanceolata 6-9 cm longa 2.5-3.5 cm lata, apice abrupte acuta basi cuneata, supra in sicco nitidula subtus glabra, nervis lateralibus 7-9-jugis supra prominentibus subtus inconspicuis, margine acriter serrata, dentis 2-4 mm remotis, petiolis 4-5 mm longis glabris. Flores terminales albi subsessiles; bracteis et sepalis 9-10 exterioribus late ovatis glabris interioribus obovatis 1.2 cm longis, extus sericeo-pilosis; petalis et staminibus non visis; ovariis 3-locularibus subglabris vel apiceum versus sparse puberulis tuberclatis, stylis 3.

Small tree, branchlets glabrous. Leaves thinly coriaceous, obovate-lanceolate, 6-9 cm long, 2.5-3.5 cm wide, apices abruptly acute, bases cuneate, slightly shiny above, glabrous below; lateral veins 7-9 pairs, protruding above in the dry state, inconspicuous below; margins sharply serrate, teeth separation 2-4 mm; petioles 4-5 mm long, glabrous. Flowers terminal, nearly sessile; bracts and sepals 9-10, outer 4 bracts broadly ovate, glabrous, others broadly obovate, longest to 1.2 cm, exterior with ash-white tomentum; petals not yet seen; stamens connate, deciduous; ovaries 3-locular, nearly glabrous or apically pubescent, tuberculate; styles 3-parted, free, glabrous.

Yunnan: Wen Shan, Pingzhai, 26 April 1960, Nanjing University Biology Dept. 15192 (holotype in PE).

4. *Camellia rhytidocarpa* Chang & Liang in Chang, Acta Sci. Nat. Univ. Sunyatseni, monogr. ser. 1: 49. 1981.

Species *C. tuberculata* valde similis, sed ramulis et foliis glabris, foliis dense et acute serratis, ovario 3-loculari, stylis 3 glabris.

Frutex, ramulis glabris. Folia coriacea oblonga 6-9.5 cm longa 2.5-3.5 cm lata, apice acuminata basi obtusa vel subrotundata, supra nitida subtus glabra, nervis lateralibus 6-7-jugis, margine serrulata, petiolis 8-12 mm longis. Flores terminales albi sessiles; bracteis et sepalis 10 interioribus 1-1.4 cm longis, sericeis; petalis 6 circ. 3.2 cm longis, basi 2/5 connatis; staminibus 2-2.5 cm longis, extimis basi connatis, tubo filamentorum 1.3 cm longo, glabro; ovariis pilosis, stylis 3 liberis 2 cm longis glabris. Capsula globosa 2-2.5 cm diam. 1-2-locularis, extus tuberculata; semina solitaria globosa brunnea.

Shrub 3 m tall, branchlets glabrous. Leaves coriaceous, oblong, 6-9.5 cm long, 2.5-3.5 cm wide, apices acuminate, bases obtuse or nearly rounded; olive-green above in the dry state, shiny; yellowish-green below, glabrous; lateral veins 6-7 pairs, margins sharply serrate; petioles 8-12 mm long, glabrous. Flowers terminal, white, sessile; bracts and sepals 10, gradually becoming large from outer to inner, outer 3 crescent shaped, inner 7 ovoid, 1-1.4 cm long, exterior sericeous; petals ca. 6, 3.2 cm long, basal 2/5 connate; free filaments and filament tube glabrous, filament tube adnate with petals; ovaries pilose; styles 3-parted, 2 cm long, glabrous. Capsules globose, 2-2.5 cm in diameter, 1-2-locular, 1 seeded; sometimes bi-globose, 2-locular, 1 seed per locule; exterior tuberculate or rugose; seeds globose, dark brown, exterior pubescent.

Guangxi: Longsheng, Tianping Shan, 19 November 1957, Tan Haofu 700908 (holotype in SCBI), 700700, 700895; Damiao Shan, Chen Shaoqing (S. C. Chun) 16727; Baise, Liang Shengye (S. Y. Liang) 25; Donglan, Guangxi Forestry Investigation Team 46427.

Hunan: Qianyang, Li Zetang 500.

Guizhou: Chishui, Beijing Bot. Inst. Bijie Team 1278.

5. *Camellia litchii* Chang, Acta Sci. Nat. Univ. Sunyatseni, monogr. ser. 1: 50. 1981.

Species C. *tuberculata* affinis, sed ramulis glabris, foliis tenuioribus ovatis basi rotundatis, nervis lateralibus pluribus 10-12-jugis, pericarpio tenui differt.

Frutex vel arbor parva, ramulis glabris. Folia tenuiter coriacea ovata vel longe ovata 6-8.5 cm longa 2.6-3.8 cm lata, apice acuminata vel acuta, basi rotundata interdum late cuneata, supra nitidula subtus glabra, nervis lateralibus 10-12-jugis in sicco supra prominentibus subtus conspicuis, margine serrata, dentis 2-3 mm remotis, petioli 4-6 mm longi, glabri. Flores non visi. Capsula globosa circ. 2 cm diam., 3-locularis, pericarpio tuberculato et irregulariter crispo 1.5 mm crasso; semina solitaria in quoque loculo semiglobosa 1.4 cm diam. brunnea extus puberula.

Shrub or small tree, branchlets glabrous. Leaves thinly coriaceous, ovate to long-ovate, 6-8.5 cm long, 2.5-3.8 cm wide, apices acuminate or acute; bases rounded, sometimes broadly cuneate or slightly cordate; green above in the dry state, slightly shiny, glabrous; light green or the same color below, glabrous; lateral veins 10-12 pairs, protruding above, faintly evident below; margins sharply serrulate, teeth every 2-3 mm; petioles 4-6 mm long, glabrous. Flowers not seen. Capsules globose, ca. 2 cm in diameter, 3-locular, 1 seed per locule; pericarp tuberculate or irregularly rugose, 1.5 mm thick; seeds semiglobose, 1.4 cm in diameter, dark brown, exterior pubescent.

Guizhou: Border with Guangxi, Pingzhou, 14 September 1930, Jiang Ying (Y. Tsiang) 7130 (holotype in PE, isotype in A).

6. *Camellia parvimuricata* Chang, Acta Sci. Nat. Univ. Sunyatseni, monogr. ser. 1: 51. 1981.

Species a C. *tuberculatae* et C. *rhytidcarpae* differt foliis minoribus oblongo-ovatis, perulis petalisque sericeis, ovario glabro.

Frutex, ramulis atro-brunneo-hirsutis. Folia coriacea elliptica vel ovato-elliptica 4-6 cm longa 1.5-2.5 cm lata, supra acuminata vel caudata, caudo circ. 1 cm longo, basi late cuneata vel subrotundata, supra in sicco olivacea-brunnea nitidula, ad costam puberula, subtus glabra, nervis lateralibus 6-7-jugis supra leviter prominentibus subtus conspicuis, margine dense serrulata, dentis 1-2 mm remotis, petiolis 3-5 mm longis pubescentibus. Flores terminales circ. 2.5 cm in diametro, pedicellis brevissimis; bracteis 6 semiorbicularibus vel orbicularibus 2-3.5 mm longis dense sericeis, sepalis 5 basi connatis late ovatis 7-8 mm longis densius sericeis; petalis 7 basi 5 mm connatis obovatis 1.2-2 cm longis sparse sericeis; staminibus 1.4 cm longis subliberis glabris; ovariis sulcatis, stylis 3-4 liberis 1 cm longis. Capsula immatura tuberculata.

Shrub, branchlets dark-brown hirsute. Leaves coriaceous, ovate-elliptic, 4.6 cm long, 1.5-2.5 cm wide, apices acuminate or caudate with a ca. 1 cm long cauda, bases broadly cuneate or slightly rounded; dark olive brown above in the dry state, slightly shiny, midveins puberulent; yellowish-brown below, glabrous; lateral veins 6-7 pairs, margins densely serrulate; petioles 3-5 mm long, pubescent. Flowers terminal, 2.5 cm in diameter, pedicels extremely short; bracts 6, semi-orbicular to orbicular, 2-3.5 mm long, apices rounded, back sericeous; sepals 5, basally slightly connate, broadly ovate, 7-8 mm long, apices rounded, back densely sericeous; petals 7, basally connate for ca. 5 mm, outer 2 obovate, middle of the back relatively thick, sericeous, margins membranous and glabrous, 1.2 cm long, 8-9 mm wide; inner 5 petals obovate, 1.8-2 cm long, apices rounded, middle of the back slightly sericeous; stamens 1.4 cm long, except for the outer connate filaments the others are free, filaments glabrous; ovaries glabrous; styles

3-4-parted, completely free, 1 cm long, stigmas slightly expanded. Immature capsules tuberculate.

Hunan: Qianyang, Erqu, Daping, Li Zetang 1926, 1729 (holotype in SCBI); Xuefeng Shan, Li Zetang 1101, 2996; Sheoyang Prefecture, Dongkou, 24 June 1959, Tan Peixian 63057; Xuefeng Shan, Xiaojiageng, Qi Chengjing 3329.

Hubei: Lichuan, Maoba, Wangjiaping, shrub 1.5 m tall, Dai Lunying and Qian Zhonghai 752.

Sichuan: Nanchuan, Jinfo Shan, Li Guofeng 63590.

19 *Camellia parvimuricata* Chang
1. flowering branch; 2. petals and stamens; 3. stamen; 4. pistil; 5. ovary cross section.

SECTION IX.
Luteoflora Chang, Acta Sci. Nat. Univ. Sunyatseni 1982(3): 73. 1982.

Flores lutei parvi axillares vel terminales sessiles; bracteae et sepala indistincta subcoriacea semipersistentes; petala basi connata inpatentia; stamina 2-seriata exteriora connata, ovaria 3-locularia, stylis subliberis brevibus, semina singularia in quoque loculo.

A sect. *Paracamellia* floribus luteis, petalis connatis inpantentibus, staminibus basi connatis differt.

Flowers yellow, small, axillary or terminal, sessile; bracts and sepals not differentiated, shiny and coriaceous, semipersistent; petals connate, not patent; stamens in 2 series, outer whorl connate; ovaries 3-locular; styles short, free; 1 seed per locule.

This section differs from section *Paracamellia* by having yellow flowers, petals connate and not patent, and stamens basally connate.

1 species, endemic to China.

Type: *Camellia luteoflora* Y. K. Li

1. ***Camellia luteoflora*** Y. K. Li in Chang, Acta Sci. Nat. Univ. Sunyatseni 1982(3): 72. 1982.

Frutex vel arboscula 1.2-5.5 m altus, corice cinereo vel atrocinereo; ramulis glabris; gemmis tubinatis dense albido-seriaceis. Folia coriacea oblonga vel elliptica 6.5-12 cm longa maxima ad 17 cm longa, 1.7-5.4 cm lata latissima 7 cm lata, apice acuminata vel abrupte acuta, basi late cuneata, costa supra impressa, nervis lateralibus utrinsecus 6-8, ut retis venularum supra impressis, margine revoluta remote serrulata, dentis irregularibus, inter se 2.5-8 mm distantibus; petiolis 6-12 mm longis, cinereo-pubescentibus. Flores solitarii axillares vel terminales parvi sessiles, lutei, 1-1.8 cm diam.; bracteae et sepala indistincta, 8-10, subcoriacea, semiorbiculata ad late elliptica, 4-10 mm longa, sparse puberula, semipersistentia; petala 7-8, lutea, 11-15 mm longa, basi 4 mm connata, inpatentia, late elliptica vel obovato-elliptica, apice retusa, glabro vel ciliata; stamina 2-seriata, 13 mm longa, exteriora basi connata, tubo staminis 8 mm longo, glabro, antheris luteis basifixis; ovario ovoidea albo-pubescenti, 3-loculari, stylo 5 mm longo, apice 3-lobato. Capsulae globosae vel ovoideae 1 cm in diametro; semina in quoque loculo singularia.

Shrub or small tree 1.2-5.5 m tall, bark brown or dark brown, branchlets glabrous; buds cone-shaped, whitish pubescent. Leaves oblong or elliptic, 6.5-12 cm long, 1.7-5.4 cm wide, largest leaves 17 x 7 cm, apices acuminate or abruptly acute, bases broadly cuneate, midveins sunken above, lateral veins 6-8 pairs and sunken above to the same extent as the reticulating veins; margins revolute, sparsely serrulate, teeth separation 2.5-8 mm; petioles 8-12 mm long, brownish pubescent. Flowers solitary, axillary or terminal, yellow, 1-1.8 cm in diameter, sessile; bracts and sepals not differentiated, 8-10, nearly coriaceous, semi-rotundate to broadly orbiculate, 4-10 mm long, sparsely puberulous, semipersistent; petals 7-8, 11-15 mm long, basal 4 mm connate, not patent at anthesis, broadly elliptic or obovate-elliptic, apices emarginate, glabrous or ciliate; stamens 2 series, 13 mm long, outer whorl connate at base, filament tube 8 mm long, glabrous; anthers yellowish, basifixed; ovaries ovoid, whitish-pubescent, 3-locular. Capsules globose or ovoid, 1 cm in diameter; seeds small, 1 seed per locule.

Guizhou: Chishui, Jinsha Commune, Hejing Production Brigade, 13 November 1981, Zeng Fanan and Yang Fanzen 8156 (holotype in Guizhou Sylviculture Institute, isotype in SYS).

Phylogenetically, *Camellia luteoflora* is closely related to *Paracamellia*, but flowers are yellow, petals connate at base, not patent in blooming, stamens connate to form a short tube. As to section *Chrysantha*, besides the yellow flowers, the other characters of *C. luteoflora* are quite different from those of *Chrysantha* which is a section of subgenus *Thea* with its bracts and sepals differentiated, coriaceous, persistent, flowers large, petals expanded, stamens and styles much larger.

Camellia luteoflora occurs in Chishui, in northwestern Guizhou at an altitude of 900-1000 m sporadically in forests or on cliffs.

SECTION X.
Camellia

Flowers large, red or occasionally white, 5-10 cm in diameter, solitary, terminal or axillary, sessile; bracts and sepals not differentiated, 10-21, forming 2-4 cm long perules, deciduous after flowering; petals 5-12, highly connate; stamens in many series, filaments of outer whorl connate into a short tube; ovaries 3-locular, rarely 5-locular, pubescent or glabrous; styles connate, rarely free, apex shallowly 3-cleft, rarely 5-cleft. Capsules large, 1-5 seeds per locule.

33 species, mostly in China, 3 species in Japan.

Type: *C. japonica* L.

KEY TO SECT. *CAMELLIA*

1. Ovaries pubescent . . . . Subsect. I. *Reticulata* Chang
   2. Outer filament whorl or filament tube pubescent. . . . . Ser. I. *Villosae* Chang
      3. Leaves large, mostly greater than 10 cm long.
         4. Leaves elliptic, 8-12 cm long; bracts and sepals 10-15.
            5. Leaf apices abruptly acute, bluntly serrate; bracts and sepals 10 . . . .
               . . . . 1. *C. omeiensis* Chang
            5. Leaf apices acute, margins sharply serrate, bracts and sepals 15 . . . .
               . . . . 2. *C. polyodonta* How ex Hu
         4. Leaves oblong, 16 cm long, apices long caudate; bracts and sepals 10, petals 6 . . . . .
            . . . . 3. *C. lapidea* Wu
      3. Leaves small, 5-9 cm long; bracts and sepals 7-14.
         6. Bracts and sepals 7-8, leaf backs glabrous.
            7. Flowers red . . . . 4a. *C. mairei* (Lév.) Melchior var. *mairei*
            7. Flowers white . . . . 4b. *C. mairei* var. *alba* Chang
         6. Bracts and sepals 14, leaf backs pubescent . . . . 5. *C. villosa* Chang & Liang
   2. Outer filament whorl and filament tube completely glabrous. . . . Ser. II. *Reticulatae* Chang
      8. Ovaries 4-5-locular, styles apically 4-5-cleft.
         9. Bracts and sepals 9-10, coriaceous, brown velutinus . . . . . 6. *C. kweichowensis* Chang
         9. Bracts and sepals 12, somewhat membranous, ash-white sericeous . . . .
            . . . . 7. *C. albovillosa* Hu ex Chang
      8. Ovaries 3-locular, styles 3-cleft.
         10. Flowers white, sometimes light red.
             11. Leaf margins completely serrate, leaves 4-9 cm long, fruit smaller than 7 cm in diameter.
                 12. Leaves ovate to ovate-oblong, bases slightly rounded.
                     13. Sepals coriaceous, pubescent, 1.4 cm long, 4-5 mm thick . . . .
                         . . . . 8. *C. albescens* Chang
                     13. Sepals membranous, nearly ciliate, 1 cm long; petals 8, valves 2-3 cm thick. . . . . 9. *C. tunganica* Chang & Lee
                 12. Leaves elliptic, bases broadly cuneate . . . . . 16c. *C. pitardii* var. *alba* Chang
             11. Upper ⅓ of leaf margins serrate, leaves longer than 15 cm long, fruit to 10 cm in diameter. . . . . 12b. *C. semiserrata* var. *albiflora* Hu & Huang
         10. Flowers red.
             14. Leaves elliptic or oblong elliptic, 2 times as long as wide.
                 15. Upper ½ to ⅓ of the leaf margins serrate.

16. Bracts and sepals 9-10.
    17. Capsules with 3-5 seeds per locule.
        18. Pericarp 3-4 cm thick, woody; fruit 10 cm in diameter, sepal pubescence yellowish-brown, seeds pubescent..10. *C. trichosperma* Chang
        18. Pericarp 1-1.5 cm thick, corky; fruit 4.5-6 cm in diameter, sepal pubescence ash-white, seeds glabrous...11. *C. phellocapsa* Chang & Lee
    17. Capsules with 1-2 seeds per locule, fruit 6-7 cm in diameter, pericarp crustaceous, seeds glabrous..2a. *C. semiserrata* Chi var. *semiserrata*
16. Bracts and sepals 21, thickly coriaceous.... 13. *C. multiperulata* Chang
15. Leaf margins completely serrate, or upper ⅔ serrate.
    19. Leaf bases rounded, bracts and sepals somewhat membranous ....14. *C. lungshenensis* Chang
    19. Leaf bases cuneate; bracts and sepals 10, coriaceous.
        20. Leaves sharply serrate.
            21. Sepals yellowish-brown tomentose, petals 5-6 cm long, leaves to 10 cm long................15. *C. reticulata* Lindl.
            21. Sepals yellowish-brown puberulent, flowers 3-4 cm long, leaves 5-7 cm long....16a. *C. pitardii* Cohen-Stuart var. *pitardii*
        20. Leaves bluntly serrate.
            22. Leaf lower surfaces not glandular punctate, sepals puberulent, flowers pink, filament tubes shorter than the free filaments, stamens in many series........17. *C. hiemalis* Nakai
            22. Leaf lower surfaces glandular punctate, sepals pubescent, flowers deep red, filament tubes longer than free filaments, stamens in 1-2 series..............18. *C. uraku* Kitamura
14. Leaves oblong or lanceolate, sometimes oblanceolate, 3-4 times as long as wide.
    23. Leaf bases obliquely cordate or rounded, sepal apices acute ....19. *C. edithae* Hance
    23. Leaf bases cuneate, sepal apices rounded.
        24. Leaf apices long caudate, cauda 1.5-2.5 cm long.
            25. Leaves serrulate, veins obscure on both surfaces; bracts and sepals 8, coriaceous; petals 5-6 ...20. *C. xylocarpa* (Hu) Chang
            25. Leaves coarsely serrate, veins conspicuous on both surfaces, bracts and sepals 9-10, petals 6-8 ....16a. *C. pitardii* Cohen-Stuart var. *pitardii*
        24. Leaf apices acute or acuminate.
            26. Leaves longer than 10 cm.
                27. Bracts and sepals membranous in the dry state; styles 3, free.........................21. *C. hongkongensis* Seem.
                27. Bracts and sepals coriaceous, 8-10; styles connate, apically shallowly 3-cleft; leaves oblanceolate or oblong; styles connate, apices 3-cleft.
                    28. Leaves oblong-oblanceolate, half serrate, lateral veins obscure, bracts and sepals 8, fruit globose ....22. *C. cryptoneura* Chang
                    28. Leaves oblong, margins completely serrate, lateral veins evident, bracts and sepals 8-11.
                        29. Leaves 10-16 cm long, sepals sericeous, pericarp corky.

30. Sepals completely pubescent, fruit ovate......
.... 23. *C. oviformis* Chang
30. Sepals only pubescent in the middle, fruit oblate
.... 24. *C. compressa* Chang & Wen
29. Leaves 9-12 cm long, sepals hirsute, pericarp hard
.... 25. *C. setiperulata* Chang & Lee
26. Leaves 5-9 cm long.
31. Leaves yellowish-green in the dry state, lateral veins conspicuous, pericarp thin, flowers light red.
32. Leaves 5-7 cm long, 1-2 cm wide, apices obtuse; flowers small, sepals nearly glabrous, branchlets pubescent
.... 26. *C. saluenensis* Stapf ex Bean
32. Leaves 6-9 cm long, 2.5-3.5 cm wide, apices acuminate; petals 3-4 cm long, sepals pubescent, branchlets glabrous or pubescent.
33. Sepals coriaceous, petals 3-3.5 cm long, branchlets pubescent ..... 17b. *C. pitardii* var. *yunnanica* Sealy
33. Sepals somewhat membranous, petals 3.5-4.5 cm long, branchlets pubescent.....................
.... 14. *C. lungshenensis* Chang
31. Leaves dark green in the dry state, veins obscure, pericarp 4-20 mm thick.
34. Leaves with 5-6 pairs of lateral veins, petals 5, pericarp 1-2 mm thick........... 20. *C. xylocarpa* (Hu) Chang
34. Leaves with 7-8 pairs of lateral veins, petals 5, pericarp 4-6 mm thick, styles glabrous....................
.... 27. *C. boreali-yunnanica* Chang
1. Ovaries glabrous, pericarp woody..................... Subsect. II. *Lucidissima* Chang
35. Bracts and sepals 14-16, seeds 8 per locule, leaf surfaces shiny.....................
.... 28. *C. chekiangoleosa* Hu
35. Bracts and sepals 7-10, seeds 1-5 per locule, leaf surfaces shiny or dull.
36. Outer filament whorl nearly free; leaves extremely shiny, over 10 cm long......
.... 29. *C. lucidissima* Chang
36. Outer filament whorl connate into a short tube; leaves somewhat shiny, variable in length.
37. Leaves elliptic, clearly serrate, bracts and sepals 9-10.
38. Leaves greater than 10 cm long, 7-10 cm wide, upper half serrate; fruit large, 10 cm wide, seeds 5 per locule ..... 30. *C. magnocarpa* (Hu & Huang) Chang
38. Leaves less than 10 cm long, 3-4 cm wide, margins completely serrate; fruit 3-4 cm wide, seeds 1-3 per locule................... 31. *C. japonica* L.
37. Leaves lanceolate or narrowly oblong, margins entire or serrulate, bracts and sepals 7-10.
39. Leaves narrowly oblong, apices acute, margins nearly entire; flowers 3.5-5.7 cm long................................ 32. *C. subintegra* Huang
39. Leaves narrowly lanceolate, apices long caudate, margins serrulate; flowers ca. 3 cm long...................... 33. *C. longicaudata* Chang & Liang

SUBSECT. I.
Reticulata Chang, Acta Sci. Nat. Univ. Sunyatseni, monogr. ser. 1: 56. 1981.

Ovarium tomentosum, pericarpium crassime coriaceum brunneum.
Ovaries tomentose; pericarp brown, pubescent and thickly coriaceous.
2 series and 27 species distributed in China and Japan.
Type: *C. reticulata* Lindl.

SER. I.
Villosae Chang, Acta Sci. Nat. Univ. Sunyatseni, monogr. ser. 1: 56. 1981.

Filamentum extimum pubescens.
Outer filament whorl pubescent.
5 species, endemic to China.
Type: *C. villosa* Chang & Liang

1. *Camellia omeiensis* Chang, Acta Sci. Nat. Univ. Sunyatseni, monogr. ser. 1: 56. 1981.

Species a. *C. mairei* differt foliis floribusque multo majoribus; a *C. lapidea* differt foliis ellipticis non caudatis, floribus majoribus; a *C. polydonta* differt foliis tenuioribus apice abrupte acutis, floribus majoribus, bracteolis et sepalis paucioribus, stylis longioribus glabris.

Frutex vel arbor parva, ramulis glabris. Folia coriacea elliptica 9-12 cm longa 4-5.5 cm lata, apice abrupte acuta, acumene obtuso, basi rotundata vel obtusa, supra nitidula subtus glabra, nervis lateralibus 6-7-jugis ut retis nervorum supra impressis subtus prominentibus, margine serrata, petiolis 1-1.5 cm longis. Flores solitarii terminales rubri 9 cm in diametro sessiles; bracteis et sepalis 10, semiorbiculatis vel orbiculatis exterioribus 3-6 mm longis 4-9 mm latis, interioribus 1.5-2 cm longis orbiculatis, sericeis, deciduis; petalis 8-9, suboribuclatis vel late obovatis, 2-5.5 cm longis, basi 1.5 cm connatis; staminibus 3.5-4 cm longis, extimis pilosis basi 1-1.5 cm connatis; ovariis brunneo-pilosis, stylis 3-3.5 cm longis glabris, apice 3-fidis, lobis circ. 1 cm longis. Capsula globosa 3-locularia, 3-valvata dehiscens, valvis crasse lognosis; semina bina in quoque loculo.

Shrub or small tree 4 m tall; branchlets glabrous, dark brown. Leaves coriaceous, elliptic, 9-12 cm long 4-5.5 cm wide, apices abruptly acute with the extreme tip slightly obtuse, bases rounded or obtuse; dark green above in the dry state, slightly shiny or dull, glabrous; brown below, glabrous; lateral veins 6-7 pairs; margins serrate; petioles 1-1.5 cm long, flattened, glabrous. Flowers terminal, sessile, red, 9 cm in diameter, 5-6 cm long; bracts and sepals 10, lower 3-4 perules semiorbicular, 3-6 mm long, 4-9 mm wide, exterior sericeous; inner perules nearly orbicular, 1.5-2 cm long, sericeous, deciduous after flowering; petals 8-9, outer 2-3 nearly orbicular, 2-3.5 cm long, back sericeous; other 6-7 petals broadly obovoid, 4-5.5 cm long, 3-4 cm wide, back glabrous, apices emarginate or rounded, basally connate for ca. 1.5 cm; stamens 3.5-4 cm long, basal half of outer whorl connate into a short tube, upper half free, pubescent; filament tube pubescent, inner filament whorl completely pubescent; ovaries brown pubescent; styles 3-3.5 cm long, glabrous, apical 1 cm 3-cleft. Capsules globose, not yet ripe, with brown pubescence, 3-locular, 2 seeds per locule, pericarp thickly woody.

Sichuan: Emei Shan, Yang Guanghui 53882 (holotype in SYS), 55133; Nanchuan, Maoping, Xiongjihua, Zhou Zilin 91288; same loc., PE Sichuan Plant Resources Investigation Team 819, elev. 1500-1800 m; same loc., elev. 1600 m, Zheng Zishan 450.

2. *Camellia polydonta* How ex Hu, Acta Phytotax. Sin. 10: 135. 1965.

Leaves elliptic, long acute, margins serrations extremely sharp. Bracts and sepals 15, fruit globose, pericarp extremely thick, seeds 3-5 per locule.

Guangxi: Lingui, Wantian, Huang Zhuojie (T. C. Huang) 2084 (holotype in SCBI); same loc., SCBI Resources Investigation Team 1021A; same loc., Feng Jinyong 1051; Liang Shengye (S. Y. Liang) s.n. (SYS no. 137640).

Liang Shengye's specimen, introduced to Nanning from the type locality has leaf blades less than 9 cm long and 4 cm wide.

Camellia omeiensis Chang
1. flowering branch; 2. petals and stamens.

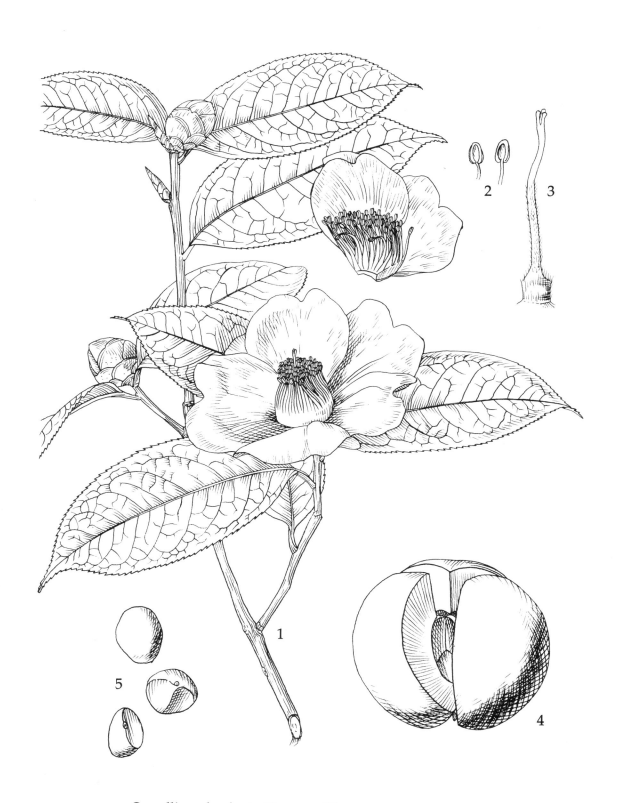

Camellia polyodonta How ex Hu
1. flowering branch; 2. anthers; 3. pistil; 4. capsule; 5. seeds; 6. petals and stamens.

3. *Camellia lapidea* Wu in Engler, Bot. Jahrb. 71: 190. 1940.
Camellia mairei var. *lapidea* (Wu) Sealy, Rev. Gen. Camellia, 174. 1958.
Leaves 16 cm long, long-caudate, lateral veins 8-9 pairs. This species differs from C. *mairei* by having very large leaves, the latter has short leaf blades, apices abruptly acute, only 5-6 pairs of lateral veins and capsules small.
Guizhou: Anlong, Jiang Ying (Y. Tsiang) 9324A.
Guangdong: Huaiji, Liu Yingguang 2706; Lian Shan, Tan Peixiang 58693.
Guangxi: Qin Xian, Chen Xhaoqing (S. C. Chen) 3914; Pingnan, Huang Zhi (C. Wang) 40720; Dayao Shan, Li Yinkun 400422; Beiliu, Zeng Bangguang 8965.

4. *Camellia mairei* (Lév.) Melchior in Engler, Nat. Pflanzenfam. ed. 2, 21: 129. 1925.
Thea mairei Lév., Sertum Yunnan 2, 1916.
Camelliastrum mairei (Lév.) Nakai, Journ. Jap. Bot. 16: 701. 1940.
Thea pitardii Rehd., Journ. Arn. Arb. 15: 99. 1934, *pro parte, auct. non* (Cohen-Stuart) Rehd.

4a. *Camellia mairei* (Lév.) Melchior var. *mairei*
This species is very similar to C. *saluenensis*, but it has branchlets glabrous, leaf apices acuminate, filaments densely pubescent. In addition to these three characteristics the specimens from Yunnan have relatively small flowers, 2-2.5 cm long.
Yunnan: Mengzi, Jiang Ying (Y. Tsiang) 13256; Dayao, Yang Jingsheng and Liu Dachang 2922; Yuanjiang, New York Botanical Garden 13558; Qiubei, Cai Xitao (H. T. Tsai) 51419; Guangxi, Wang Qiwu (C. W. Wang) 83990, 86350.
Guangxi: Lingle, Cao Shan, Zhang Zhaoqian 10304.
Sichuan: Nanchuan, Li Guofeng 55120; same loc., Guang Kejian 1222.
The Sichuan specimens have peach-blossom-red flowers, leaf blades relatively narrow, apices long pointed. Obviously they are transitional to var. *alba*.

4b. *Camellia mairei* var. *alba* Chang, Acta Sci. Nat. Univ. Sunyatseni, monogr. ser. 1: 60. 1981.
Foliis oblongis, floribus albis bracteolis et sepalis 8, petalis 7, staminibus pilosis, ovario piloso 3-locularibus stylis glabris.
Leaves oblong, lateral veins 5-7 pairs. Flowers white, bracts and sepals 8, petals 7, outer whorl of filaments pubescent; ovaries 3-locular, pubescent; styles glabrous, 3-cleft.
Sichuan: Emei Shan, Chudian, right side, elev. 1650, 18 November 1964, Guan Kejian 2760 (holotype in PE).

5. *Camellia villosa* Chang, Acta Sci. Nat. Univ. Sunyatseni, monogr. ser. 1: 60. 1981.
A C. *polyodonta* differt foliis minoribus subtus villosis, floribus brevioribus 3.5-4 cm longis, capsulis minoribus 4 cm diametro.
Frutex vel arbor parva, ramulis glabris. Folia coriacea oblonga 7-9.5 cm longa 3-4 cm lata, apice abrupte acuta, basi rotundata vel obtusa, supra nitidula subtus sparse pilosa, nervis lateralibus 7-9-jugis in sicco supra impressis, margine superiore ⅔ serrulata, petiolis 5-7 mm longis pilosis. Flores rubri 6-7.5 cm diam.; bracteis et sepalis 14 ovoideis maximis 1.3 cm longis sericeis; petalis 7 obovatis 3-3.5 cm longis extus pubescentibus; staminibus 4-seriatis 2.5 cm longis, tubo filamentorum pubescente; ovariis pilosis 3-locularibus, stylis 1.5 cm longis apice 3-fidis; ovula 5-6 in quoque loculo. Capsula globosa 5 cm in diametro 3-valvata dehiscens, valvis 5 mm crassis; semina 3-4 in quoque loculo, 1.5-2 cm longa.
Shrub or small tree, branchlets glabrous, old branches greyish-white. Leaves coriaceous, oblong 7-9.5 cm long, 3-4 cm wide, apices acute with an 8-10 mm long cauda, bases rounded or obtuse; brown above in the dry state, glabrous, slightly shiny;

1-3. *Camellia lapidea* Wu   4-5. *Camellia longicaudata* Chang & Liang
1. flowering branch; 2. petal and stamens; 3. pistil; 4. flowering branch;
5. pistil.

23 *Camellia mairei* (Lév.) Melch.
1. flowering branch; 2. petal; 3. petal and stamens; 4. pistil.

same color below, villous; lateral veins 7-9 pairs, perpendicular to the midvein; basal ⅓ serrate, petioles 5-7 mm long, upper surface glabrous, lower surface pubescent. Flowers red, sessile; bracts and sepals 14, ovate, to 1.3 cm long, outer surfaces pubescent; stamens in 4 series, 2.5 cm long, filament tube pubescent, inner free filaments pubescent; ovaries pubescent, 3-locular, each locule with 5-6 ovules; styles 1.5 cm long, apices deeply 3-cleft. Capsules globose, 5 cm in diameter, brown, villous, 3-locular, 3-4 seeds per locule, 3-valvate dehiscent, valves 5 mm thick, bracts and sepals not persistent, pedicels extremely short; seeds 1.5-2 cm long, brown.

Hunan: Xuefeng Shan, Li Zetang 1874.

Guangxi: Longsheng, Luan Shan, 15 October 1956, Huang Zuojie (T. C. Huang) 2060 (holotype in SCBI); Dayao Shan, Zhongliang Commune, Lu Qinghua 4377; Damiao Shan, Shen Shaoqing (S. C. Chun) 15205, 17234; He Xian, Huang Zhuomin s.n. (Guangxi For. Inst. no. 12257); Guchen, Li Binggui s.n. (HNTC no. 7294).

Guizhou: Rongjiang, Yueliang, Bot. Inst. Acad. Sin. Qiannan Team 2953, 4001.

This species has 14 perules, 7 petals, and is very close to *C. polyodonta*, except the leaves are shorter, back surface is pubescent, and the flowers and fruit are relatively small.

SER. II.
Reticulatae Chang, Acta Sci. Nat. Univ. Sunyatseni, monogr. ser. 1: 61. 1981.

Tubo filamentorum et filamentis liberis extimis glabris, ovariis pilosis, stylis connatis.
Outer filament tube and free filaments glabrous, ovaries pilose, styles united.
22 species.
Type: *C. reticulata* Lindl.

6. *Camellia kweichowensis* Chang, Acta Sci. Nat. Univ. Sunyatseni, monogr. ser. 1: 61. 1981.

Species foliis oblongis, bracteis sepalisque 9-10 coriaceis, petalis 9 circ. 6 cm longis, staminibus 2-2.5 cm longis, tubo 5 mm longo, glabris, ovario 5-loculi, stylo 5-fido, capsulis oblatis 5-locularibus distincta.

Arbor parva, ramulis glabris. Folia coriacea oblonga 6-10.5 cm longa 3-4 cm lata, apice subacuta vel acuminata, basi late cuneata vel subrotundata, nitida glabra, nervis lateralibus 6-7-jugis, margine serrulata, petiolis 8-12 mm longis. Flores rubi 10 cm in diametro sessiles; bracteis et sepalis 9-10 coriaceis semiorbiculatis vel late ovatis 5-17 mm longis extus flavido-pubescentibus; petalis 9 obovatis inaequalibus 3-6 cm longis, basi 1-1.5 cm connatis; ovariis 5-locularibus pilosis, stylis 2 cm longis glabris apice 5-fidis. Capsula compresse globosa (immatura) 3 cm diametro, 1.8 cm alta, leviter 5-sulcata; semina 1-2 in quoque loculo.

Small tree. Leaves oblong, 6-10.5 cm long, 3-4 cm wide. Flowers red, 10 cm in diameter; bracts and sepals 9-10, coriaceous, semiorbicular to broadly ovate, 5-17 mm long, ash-yellow pubescent; petals 9, outer 3 petals 3-3.5 cm long, pubescent, inner 6 petals 5-6 cm long, obovate, basally connate for 1-1.5 cm; stamens 2-2.5 cm long, glabrous, filament tube short, 5 mm long; ovaries 5-locular, pilose; styles 2 cm long, apically 5-cleft, glabrous. Capsules oblate, 5-locular, 1-2 seeds per locule, pericarp 4 mm thick.

Guizhou: Qingzhen, Jiulong Shan, 27 April 1936, Deng Shiwei (S. W. Teng) 90180 (holotype in SCBI); Qingzhen, Dazhou, 24 June 1935, Deng Shiwei (S. W. Teng) 693.

The fact that this species has 5-locular ovaries indicates that it is a comparatively primitive member of section *Camellia* and is closely related to subgenus *Protocamellia*.

24 *Camellia kweichowensis* Chang
1. flowering branch; 2. capsule.

Camellia albovillosa Hu ex Chang
1. flowering branch; 2. petal; 3. stamens; 4. stamen; 5. pistil.

7. *Camellia albovillosa* Hu ex Chang, Acta Sci. Nat. Univ. Sunyatseni, monogr. ser. 1: 62. 1981.

Ramulis pilosis, foliis ellipticis subtus pubescentibus, floribus rubris, bracteis et sepalis 12, petalis 7-8 albovillosis 3.5 cm longis, staminibus glabris distincta.

Frutex circ. 1-2 m altus, ramulis pubescentibus. Folia coriacea elliptica 6-10 cm longa 3-4 cm lata, apice abrupte acuta basi subrotundata, supra in sicco nitidula, subtus primo pubescentia demum glabrescentia, nervis lateralibus 8-10-jugis, margine serrulata, petiolis 7-8 mm longis. Flores terminales rubri sessiles 6-7 cm in diametro; bracteis et sepalis 12 scariosis 4-12 mm longis griseo-sericeis; petalis 7-8 obovatis 3-3.5 cm longis exterioribus pubescentibus; staminibus 2-2.5 cm longis, extimis basi connatis glabris; ovariis pilosis, stylis 1-1.5 cm longis apice 3-5-fidis glabris. Fructus ignotus.

Shrub 1-2 m tall, branchlets pubescent. Leaves coriaceous, elliptic, 6-10 cm long, 3-4 cm wide, apices abruptly acute, bases nearly rounded, slightly glossy above, at first pubescent below, lateral veins 8-10 pairs, margins serrulate, petioles 7-8 mm long. Flowers terminal, sessile, red; bracts and sepals 12, membranous when dry, 4-12 mm long, greyish-white pubescent; petals 7-8, obovate, 3-3.5 cm long, exterior of outer 3-4 petals pubescent; stamens 2-2.5 cm long; outer filament whorl forming a short tube, glabrous; ovaries pubescent, 3-5 locular; styles 1-1.5 cm long, 3-5-cleft, glabrous.

Yunnan: Yuanjiang, Wuqu, Wudie, elev. 1200 m, flowers red, 23 November 1964, Li Yanhui 5988 (holotype in YNTBI); same loc., Yin Wenqing 1982.

This species is distinct by having 12 perules, membranous when dry, greyish-white pubescent; ovaries 3-5 locular; leaves basally nearly rounded, somewhat pubescent, lateral veins to 10 pairs; branchlets pubescent.

8. *Camellia albescens* Chang, Acta Sci. Nat. Univ. Sunyatseni, monogr. ser. 1: 63. 1981.

A *C. reticulata* differt foliis minoribus 4-6.5 cm longis ovatis basi rotundatis brevius petiolatis, floribus minoribus 3-3.5 cm longis albescentibus, petalis 12, antheris majoribus 3 mm longis, capsulis minoribus 2.5-3 cm diam.

Frutex circ. 1.5 cm altus, ramulis glabris. Folia coriacea ovata 4-6.5 cm longa 2.5-4 cm lata, apice subacuta basi rotundata, supra nitida subtus in sicco flavo-virentia, ultimum rubro-brunnea, nervis lateralibus circ. 6-jugis, margine serrata, petiolis 3-5 mm longis. Flores axillares albi 3-3.5 cm diam. bracteis et sepalis 10, maximis 1.4 cm longis suborbiculatis extus fluvo-brunneo-sericeis; petalis obovatis 12 glabris; staminibus 2 cm longis, extimis basi connatis glabris, antheris 3 mm longis; ovariis fluvo-pilosis, stylis 2 cm longis glabris apice 3-fidis. Capsula globosa 2.5-3 cm in diametro, 3-valvata dehiscens, valvis 4-5 mm crassis lignosis; semina 1-2 in quoque loculo.

Shrub 1.5 m tall, branchlets glabrous. Leaves coriaceous, ovate, 4-6.5 cm long, 2.5-4 cm wide, apices subacute, bases rounded, shiny above; yellowish-green below, becoming reddish-brown in the dry state; lateral veins 6 pairs, margins serrate, petioles 3-5 mm long. Flowers axillary, 3-3.5 cm long, white, subsessile; bracts and sepals 10, to 1.4 cm long, yellowish-brown sericeous; petals 12, obovate-orbicular; stamens 2 cm long, outer filament whorl forming a short tube, glabrous, anthers 3 mm long, ovaries yellow-velutinous; styles 2 cm long, glabrous, apically 3-cleft. Capsules globose, 2.5-3 cm in diameter, 3-valvate dehiscent, 1-2 seeds per locule, valves 4-5 mm thick.

Yunnan: Songming, Shaojiao, Guodong, flowers white, 6 April 1957, Qiu Bingyun 54208 (holotype in KUN); same loc., Guodong, Mao Pingyi (P.I. Mao) 135; Dayao, Liuchuan Commune, Pingfengshan Team, elev. 1900 m, Zhai Ping 60-11.

9. *Camellia tunganica* Chang & Lee in Chang, Acta Sci. Nat. Univ. Sunyatseni, monogr. ser. 1: 64. 1981.

Ramulis glabris, foliis ovato-oblongis 6-8 cm longis 2.5-3 cm latis, apice acuminatis basi subrotundatis, nervis lateralibus 5-6-jugis inconspicuis, serrulatis, petiolis 6-7 mm longis; floribus albis, 1-2 terminalibus subsessilibus, perulis 9 membranaceis lato-obovatis interioribus circ. 1 cm longis subglabris vel puberulis, petalis 8 obovatis

Camellia tunganica Chang & Lee 26 flowering branch.

emarginatis 3-4 cm longis, staminibus 1.5-2 cm longis 4-5-seriatis extimis basi connatis glabris, ovario tomentoso, stylo staminibus paulo longiore glabro apice 3-fido; capsulis globosis 4.5-5.5 cm diam. 3-locularibus, seminibus 1-2 in quoque loculo.

Small tree, branchlets glabrous. Leaves coriaceous, ovate-oblong, 6-8 cm long, 2.5-3 cm wide, apices acute, bases subrotund, shiny above; lateral veins 5-6 pairs, obscure; margins serrulate, petioles 6-7 mm long. Flowers white, 1-2 terminal, sessile; bracts and sepals 9, membranous, broadly obovate, to 1 cm long, exterior glabrous or puberulent; petals 8, 3-4 cm long, exterior greyish-white sericeous; stamens 1.5-2 cm long, outer filament whorl connate, glabrous; ovaries tomentose; styles slightly longer than stamens, glabrous, apically 3-cleft. Capsules subglobose, 4.5-5.5 cm in diameter, 3-locular, valves 2-3 mm thick, 1-2 seeds per locule.

Hunan: Dongan, Damaokou, elev. 1100 m, forest in the mountains, 16 March 1965, Xiao Zuncun 65002 (holotype in HNTC).

This species is distinct by having leaves ovate-oblong, bases subrotundate, flowers white, bracts scarious; valves of capsule thin.

10. *Camellia trichosperma* Chang, Acta Sci. Nat. Univ. Sunyatseni, monogr. ser. 1: 64. 1981.

A *C. semiserrata* sepalis fulvopilosis, capsulis majoribus 12-15 cm diam. 4-5-locularibus, valvis crassissimis 3-4 cm crassis, seminibus pilosis 3-5 in quoque loculo differt.

Arbor circ. 15 m alta, cortice albo, ramulis glabris. Folia coriacea elliptica 11-13 cm longa 4.5-5.5 cm lata, apice abrupte acuta basi cuneata, supra in sicco flavo-virentia opaca, subtus flavido-viridia glabra, nervis lateralibus 5-6-jugis utrinque conspicuis, margine superiore serrulata, petiolis 1-1.3 cm longis. Flores rubri terminales sessiles; bracteis et sepalis 9-10 late ovatis 1.5 cm longis extus flavo-brunneo-sericeis. Capsula globosa 12-15 cm in diametro, 4-5-valvata dehiscens, valvis 3-4 cm crassis pubescentibus, columella robusta 4-5-angulata; semina 3-5 in quoque loculo, puberula.

Tree 15 m tall; bark smooth, greyish-white; branchlets glabrous. Leaves coriaceous, elliptic, 11-13 cm long, bases broadly cuneate; yellowish-green above in the dry state, not shiny; lighter below; lateral veins 5-6 pairs, conspicuous on both surfaces; reticulate veins obscure, upper half of margins serrulate, petioles 1-1.3 cm long. Flowers red, terminal, sessile; bracts and sepals 9-10, broadly ovoid, 1.5 cm long, outer surfaces yellowish-brown sericeous. Capsules large, globose, 12-15 cm in diameter, 4-5-valvate dehiscent, valves 3-4 cm thick, pubescent, 3-5 seeds per locule, seeds pubescent; columella large, 4-5 angled.

Jiangxi: Xunwu, Zhonghe Commune, Waqiao Production Brigade, Jiangxi Communism University Forestry Dept. 744011 (holotype in Univ. Communism Jiangxi).

11. *Camellia phellocapsa* Chang & Lee in Chang, Acta Sci. Nat. Univ. Sunyatseni, monogr. ser. 1: 65. 1981.

Ramuli glabri. Folia coriacea elliptica 9-12 cm longa 4-6 cm lata, apice abrupte acuta basi late cuneate vel interdum subrotundata, nervis lateralibus 6-7-jugis, margine ½ subperiore calloso-serrata, petioli 1-1.5 cm longi. Flores rubelli 4-6 cm diam.; perules 9-10 coriaceae 1.8 cm longae albido-pubescentes; petalis 6-8; stamina 5-6-seriata extima connata glabra ovaria tomentosa 3-locularia, pericarpio 1-1.5 cm crasso suberifero, seminibus 4-5 in quoque loculo.

Small tree, branchlets glabrous. Leaves coriaceous, elliptic, 9-12 cm long, 4-6 cm wide, apices acute, bases broadly cuneate or subrotund, lateral veins 6-7 pairs, upper half of margin serrate, petioles 1-1.5 cm long. Flowers light red, terminal, sessile, 4-6 cm in diameter; bracts and sepals 9-10, broadly obovate, coriaceous, to 1.8 cm long, grayish-white pubescent; petals 6-8, basally connate, 2-3 cm long; stamens in 5-6 series,

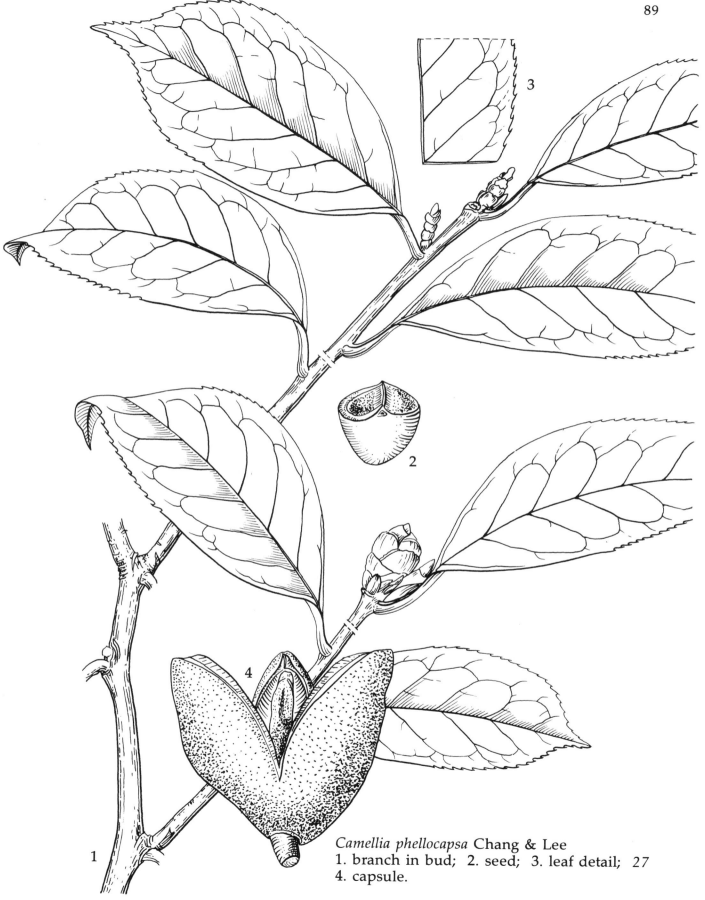

Camellia phellocapsa Chang & Lee
1. branch in bud; 2. seed; 3. leaf detail; 4. capsule.

outer filament whorl connate into a short tube, glabrous; ovaries pubescent, styles 3-cleft. Capsules pyriform, 6-8 cm long, 4.5-6 cm wide, 3-locular, 4-5 seeds per locule, valves woody, 1-1.5 cm thick.

Hunan: Chaling, Yantang Commune, elev. 550 m, mountainous region, cultivated, October 1977, Hunan For. Inst. 5678 (holotype in SYS).

This species is close to *C. semiserrata*, only the capsules are pyriform, pericarp woody, seeds 4-5 per locule. Of the members of subsection *Reticulata* with pubescent ovaries, only this species and *C. polyodonta* are 5 seeded.

12. *Camellia semiserrata* Chi, Sunyatsenia 7: 15. 1948.

12a. *Camellia semiserrata* Chi var. *semiserrata*

Leaves elliptic, 5-6.5 cm wide, only the apical ⅓ serrate. Bracts and sepals 10-11; petals 6-7, red. Capsules 4-8 cm in diameter, fruit crustaceous and thick.

Guangdong: Dinghu Shan, Li Qijing 196; Guangning, SYS 137184; Guangdong Forestry Institute, cultivated, Zhang Hongda (H. T. Chang) s.n. (SYS no. 120623); Guangning, Liang Xiangyue (H. Y. Liang) 61842 (lectotype, here designated in SCBI); Yangchun, Huang Zhi (C. Wang) 38584.

Guangxi: no loc., Liang Shengye (S. Y. Liang) 110.

12b. *Camellia semiserrata* var. *albiflora* Hu & Huang in Hu, Acta Phytotax. Sin. 10: 137. 1965.

Leaves oblong, 10-17 cm long, 4-6 cm wide; flowers white, fruit larger than the species.

Guangdong: Fengkai, Chen Shaoqing (S. H. Chun) 18468 (holotype in SCBI); same loc., Zhang Bingming 5048.

13. *Camellia multiperulata* Chang, Acta Sci. Nat. Univ. Sunyatseni, monogr. ser. 1: 66. 1981.

Species *C. semiserratae* affinis, a qua differt foliis ellipticis basi rotundatis, bracteis sepalisque pluribus cir. 19-21 crasse coriaceis majoribus 3 cm longis, petalis pluribus circ. 8-9.

Arbor, ramulis glabris. Folia crasse coriacea elliptica 10-12 cm longa 4.5-5.7 cm lata, apice abrupte acuta basi rotundata vel obtusa, supra nitida subtus glabra, nervis lateralibus 7-8-jugis in sicco utrinque prominentibus, margine superiore ⅓ acriter serrata, dentis 3-5 mm remotis circ. 1 mm longis, petiolis 1 cm longis glabris. Flores terminales solitarii rubri sessiles; bracteis et sepalis 19-21, exterioribus 4 semiorbiculatis 3-6 mm longis 5-10 mm latis glabris intus puberulis, interioribus 15-17 orbiculatis vel flabellatis 1.3-3 cm longis 1.5-3.7 cm latis glabris, coriaceis; petalis 8-9, 4-4.5 cm longis, 3-4 cm latis, obovatis basi circ. 1.2 cm connatis; staminibus 4 cm longis glabris, extimis basi connatis, tubo filamentorum 2.5 cm longo; ovariis pilosis, stylis 4 cm longis glabris, apice 3-fidis.

Small tree, branchlets glabrous. Leaves thickly coriaceous, elliptic 10-12 cm long, 4.5-5.7 cm wide, apices acute, bases rounded or obtuse; dark green above in the dry state, shiny; glabrous below; lateral veins 7-8 pairs, protruding on both surfaces in the dry state; upper half of margins sharply serrate, teeth separation 3-5 mm, teeth ca. 1 mm long; petioles 1 cm long, glabrous. Flowers solitary, terminal, often accompanying the terminal bud, red, sessile; bracts and sepals 19-21, connate into 3.5 cm long perules, exterior glabrous or basally pubescent, inner margins sericeous; outer 4 perules semi-orbicular, 3-6 mm long, 5-10 mm wide; inner 15-17 perules orbicular to flabellate-obovate, 1.3-3 cm long 1.5-3.7 cm wide, thickly coriaceous; petals 8-9, 4-4.5 cm long, 3-4 cm wide, obovoid, basally connate for ca. 1.2 cm; outer 2-3 petals shorter, exterior pubescent; stamens ca. 4 cm long, in many series, outer filament whorl basally connate, filament tube 2.5 cm long, glabrous; ovaries pilose; styles 4 cm long, glabrous, apex

1. *Camellia multiperulata* Chang
2. *Camellia lungshenensis* Chang
1-2. flowering branches.

shallowly 3-cleft, clefts 2 mm long. Capsules oblate, with 5 obtuse ridges.

Guangdong: Fengkai, Nanfeng Commune, Lishui Production Brigade, tree 5 m tall, Chen Shaoqing (S. C. Chun) 18469 (holotype in SCBI).

Guangxi: Cangwu, 8 August 1963, Huang Zuojie (T. C. Huang) s.n. (GXFI no. 3109).

This species differs from *C. semiserrata* by having leaves broadly elliptic, basally rounded; bracts and sepals 19-21, thick and hard, nearly 3 cm long; petals slightly more, 8-9.

According to the collection data the fruit is oblate, with 5 obtuse ridges, weight to 0.9 kg, the largest tree produces 60 kg of seeds per year, after drying each kg of seed produces 0.6-0.7 kg of oil.

14. ***Camellia lungshenensis*** Chang, Acta Sci. Nat. Univ. Sunyatseni, monogr ser. 1: 67. 1981.

A *C. polyodonta* foliis ovatis, floribus minoribus, bracteis sepalisque minoribus, filamentis glabris, stylis 3 liberis differt.

Arbor parva circ. 5 m alta, ramulis glabris. Folia coriacea ovata vel elliptica 7-10 cm longa 3-5.5 cm lata, apice acuta basi rotundata vel obtusa inaequilateralia, supra nitida subtus flavo-virentia glabra, nervis lateralibus 6-7-jugis utrinque inconspicuis, margine serrulata, dentis 1-2 mm remotis, petiolis 7-10 mm longis. Flores solitarii terminales coccinei 8-9 cm in diametro sessiles; bracteis et sepalis 8, semilunatis ad late obovatis sericeis, exterioribus 2-5 mm longis 3-8 mm latis, interioribus 1.2-2.2 cm longis; petalis 6 obovatis 3.5-4.5 cm longis 2-3 cm latis basi 5-7 mm connatis subglabris; staminibus 2-2.5 cm longis 3-4-seriatis, extimis basi connatis, tubo filamentorum 1-1.3 cm longis glabris; ovariis pilosis, stylis 3 liberis 1.5 cm longis glabris. Capsula globosa 3-3.5 cm in diametro, pericarpio 5-6 mm crasso piloso brunneo.

Small tree 5 m tall, branchlets glabrous. Leaves coriaceous, ovoid or elliptic, 7-10 cm long, 3-5.5 cm wide, apices acute, bases rounded or obtuse, sides not quite equal; deep green above in the dry state, shiny; yellowish-green below, glabrous; lateral veins 6-7 pairs, obscure on both surfaces; margins serrulate; petioles 7-10 mm long, glabrous. Flowers solitary, terminal, peach-blossom-red, 8-9 cm in diameter, sessile; bracts and sepals 8, outer 2-3 crescent shaped, 2-5 mm long, 3-8 mm wide, with a midrib, exterior sericeous; inner 5 perules obovoid, 1.2-2.2 cm long, pubescent; petals 6, obovoid, 3.5-4.5 cm long, 2-3 cm wide, basally connate for 5-7 mm, glabrous or apically pubescent; stamens 2-2.5 cm long, in 3-4 series, lower half of outer filament whorl connate, filament tube 1-1.3 cm long, basal 4-5 mm adnate with petals, free filaments glabrous; ovaries pubescent; styles 3-parted, ca. 1.5 cm long, glabrous.

Guangxi: Longsheng, Hongmaoyong, tree 5 m tall, flowers peach-blossom-red, 9 March 1955, Zhong Jixin 91099 (holotype in SCBI); Longsheng, Huaping, flowers pink, 10 April 1962, Yuan Shufen and Lou Lifang 5153; Longsheng, SCIB Guangfu Forest District Investigation Team 117.

This species has peach-blossom-red or pink flowers and completely free styles which differentiates it from other members of section *Camellia*. Aside from this species the only other member of this section with free styles is *C. hongkongensis*.

15. ***Camellia reticulata*** Lindl., Bot. Reg. 13: sub pl. 1078. 1827.

Desmitus reticulata (Lindl.) Raf., Sylva Tellur., 139. 1838.

Thea reticulata (Lindl.) Kochs in Engler, Bot. Jahrb. 27: 595. 1900.

Camellia heterophylla Hu, Bull. Fan Mem. Inst. Biol. Bot. 8: 37. 1937.

Leaves elliptic, 10 cm long, serrate, sometimes slightly obtuse. Bracts and sepals 10; petals 5-7, 5-6 cm long.

Yunnan: Zhu Taoping 535; Zhang Yingbo 105; G. Forrest 9596, 12133; Liu Shenno (S. N. Liu) 15313; Tengchong, Feng Guomei (K. M. Feng) 30002-30048; Zhang Hongda (H. T. Chang) 5115, 5116, 5137.

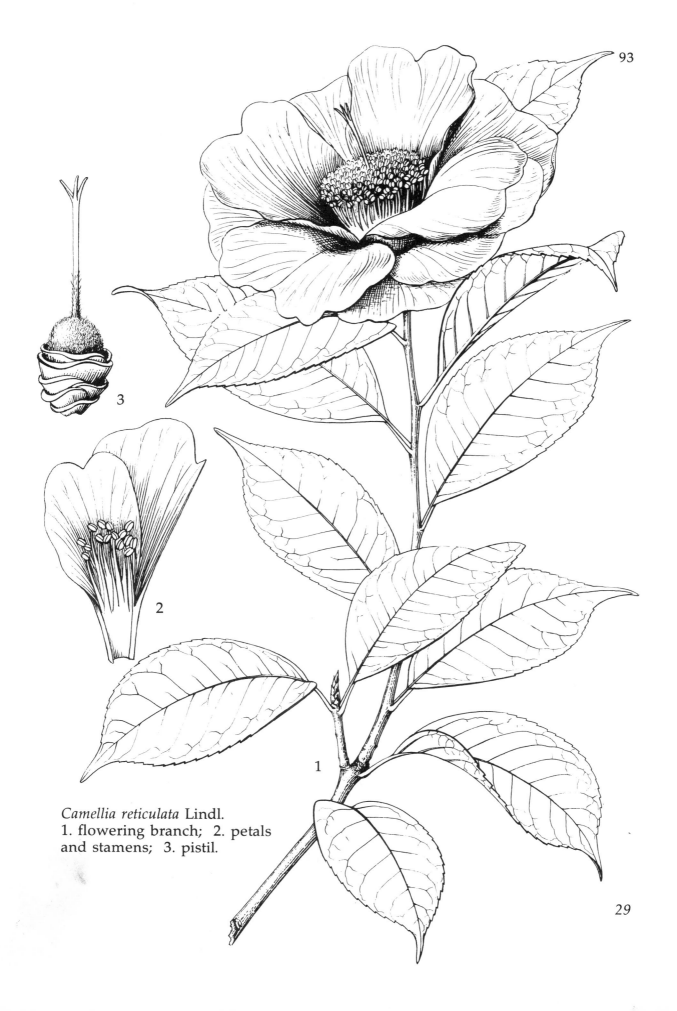

Camellia reticulata Lindl.
1. flowering branch; 2. petals and stamens; 3. pistil.

Because of the narrow leaves and light red flowers of some wild forms of this species, they are confused with *C. pitardii*. The latter has comparatively smaller or narrower leaves with sharper serrations and shorter flowers.

C. heterophylla is a cultivated variety of *C. reticulata*. Rounded leaf forms are often seen in cultivated varieties as can be seen in Zhang Yingbo 105.

This species has a long history of cultivation, and many cultivars have been maintained by vegetative propagation. Seedlings grown from these cultivars often revert to the original form.

16. ***Camellia pitardii*** Cohen-Stuart, Meded. Proefst. Thee 40: 68. 1916.
    Camellia japonica Lév., Fl. Kouy-Tcheou, 414. 1915, *pro parte, auct. non* L.
    Thea cavaleriana Lév., Cat. Pl. Yunnan, 271. 1917.
    Thea pitardii (Cohen-Stuart) Rehd., Journ. Arn. Arb. 5: 238. 1924.
    Thea grijsii Hand.-Mazz., Symb. Sinicae 7: 393. 1931, *auct. non* (Hance) O. Kuntze.
    Camellia cavaleriana (Lév.) Nakai, Journ. Jap. Bot. 14: 692. 1940.
    Camellia pitardii f. *cavaleriana* (Lév.) Sealy, Rev. Gen. Camellia, 187. 1958.

16a. ***Camellia pitardii*** Cohen-Stuart var. ***pitardii***
    Guizhou: Zunyl, A. N. Steward and Jiao Qijuan 139; Xingren, Guiyang Work Station 001; Anlong, Longlou Shan, Guiyang Work Station 72, 75, 85, 96, 108, 114, 122, 126.
    Yunnan: Mengzi, Jiang Ying (Y. Tsiang) 13218 (cultivated).
    Sichuan: Emei Shan, Fang Wenpei (W. P. Fang) 6562, 7571, 15723; same loc., Yang Yabin 67; same loc., Yang Guanghui 48540, 50081, 50202, 51455, 52443, 53951, 53958, 54282, 55039, 55474, 55536, 57334.
    Hunan: Qianyang, Li Zetang 1270.
    Guangxi: Ziyuan, Huang Deyuan 61030; Longsheng, Guangfu Forestry Investigation Team 20, 405; same loc., Tan Haofu and Li Zhongdi 70449; Xingan, Yu Shaolin 900166.

The type specimen of this species was collected from Pingfa of Guizhou (Cavalerie & Fortunat 2261), leaves elliptic, apices acute, margins sharply serrated. Formerly G. Forrest 423, collected from Chuxiong of Yunnan, was used as the type specimen and this caused the confusion in the literature. Many specimens that are the same as Forrest 423, i.e. what Sealy named as *C. pitardii* var. *yunnanica*, as seen from a large number of specimens have the characteristics of: leaves elliptic, 6-8 cm long, perules 9-10, perianth parts 6-8; capsule 3 cm in diameter, valves 3-5 mm thick. Some plants have larger flowers, deeper flower color, capsules 4-5 cm in diameter, valves 7-8 mm thick.

16b. ***Camellia pitardii*** var. ***yunnanica*** Sealy, Rev. Gen. Camellia, 187. 1958.
    Yunnan: Mengzi, A. Henry 9109; Luquan, Yin Wenqing 04, 10, 34; Kunming, E. E. Maire 1945; Dali, Wang Hanchen (H. C. Wang) 4209; Liu Shenno (S. N. Liu) 15321; Feng Guomei (K. M. Feng) 21613; Songming, Qiu Bingyun 51489; Kunming, Gao Yiming 56302.
    Sichuan: Huili, Wu Sugong (S. K. Wu) 788; Xi Changzu 514.
    Guizhou: Qianbei, Inst. of Botany Huajie Brigade 868.

This variety is commonly seen and widely distributed in Yunnan. The sepals are covered with brown hair, which is the same as *C. pitardii*, but the branchlets are often pubescent and leaves lanceolate which is similar to *C. saluenensis;* except the latter is often a small shrub, and this variety is often a small tree.

16c. ***Camellia pitardii*** var. ***alba*** Chang, Acta. Sci. Nat. Univ. Sunyatseni, monogr. ser. 1: 69. 1981.
    A typo floribus albis differt.
    Like the typical species but with white flowers.
    Hunan: Baojing, Boji Commune, Wen Xuankai 206 (holotype in SYS), 301, 306.

17. *Camellia hiemalis* Nakai, Journ. Jap. Bot. 16: 695. 1940.

Small tree, branchlets glabrous. Leaves coriaceous, elliptic or oblong, 6-9 cm long, 3-4 cm wide, lateral veins 7 pairs, serrulate. Flowers pink or white, sessile, 5 cm in diameter; bracts and sepals 8-9, broadly ovoid, 4-15 mm long, puberulent; petals 7, 3-3.5 cm long; stamens in 3-4 series, outer filament whorl forming a short tube, glabrous; ovaries pubescent, 3-locular; styles 3 cm long, apex shallowly 3-cleft.

Zhejiang: Hangzhou, Hangzhou University campus, Zheng Zhaozong 20211; HZBG 1822; Hangzhou, Gu Shan, next to Fanghe Ting, He Xianyu (H. Y. Ho) 63.

This species is a *Penjing* (bonsai) type plant introduced to Japan from Shanghai. Its appearance is similar to *C. japonica*, but the ovaries are pubescent and the flowers are pink or white.

18. *Camellia uraku* Kitamura, Acta Phytotax. et Geobot. Kyoto 14: 117. 1952.

Thea reticulata var. *rosea* Mak., Bot. Mag. Tokyo 24: 78. 1910.

Camellia reticulata var. *rosea* (Mak.) Mak., Journ. Jap. Bot. 1: 39. 1918.

Shrub, branchlets glabrous. Leaves elliptic, 7-10 cm long, 3.5-4 cm wide, brown-punctate below, lateral veins 5-7 pairs. Flowers red, glabrous, 6.5 cm in diameter; bracts and sepals 9-10, to 1 cm long, pubescent; petals 6-7, 2.8-3.6 cm long, basally connate; stamens 2 cm long, in 1-2 series; filament tube 7-8 mm long, glabrous, ovaries pubescent, styles 1 cm long.

Japan: Hondo 21 March 1915, K. Hisanti, s.n. (SYS no. 95115).

19. *Camellia edithae* Hance, Ann. Sci. Nat. Paris, ser. 4, 15: 221. 1861.

Thea edithae (Hance) O. Kuntze, Rev. Gen. Pl., 65. 1891.

Branchlets with long threadlike hairs. Leaves lanceolate; bases cordate, amplexicaul, somewhat rounded. Sepals ovate, apices acute; ovaries pubescent, styles deeply 3-cleft. Capsules 2 cm in diameter, globose, 3-locular, 1-2 seeds per locule, pericarp thin, bracts not persistent.

Jiangxi: Xunwu, Yang Xianxue (C. X. Yang) 12654; Shi Xinghua 760232.

Fujian: Tang Ruiyong 608.

Guangdong: Pingyuan, Deng Liang 4476; Fengshun, Qingliang Shan, F. A. McClure 422 (C.C.C. 6761); Dabu, Zeng Huaide (W. T. Tsang) 21050, 21101, 21632.

20. *Camellia xylocarpa* (Hu) Chang, Acta Sci. Nat. Univ. Sunyatseni, monogr. ser. 1: 71. 1981.

Yunnanea xylocarpa Hu, Acta Phytotax. Sin. 5: 280. 1956.

From the appearance of the flower there is no doubt that this species belongs to subgen. *Camellia* sect. *Camellia*. It blooms between February and March, but the fruit is not yet mature in May; pericarp thick and woody, 1-2 cm thick; usually 3-locular; with remnants of 10 bracts and sepals, to 1.4 cm long, brown pubescent; leaves small, lanceolate or oblong, lateral veins obscure.

Yunnan: Fengqing (Shunning), elev. 2400 m, small tree 6 m tall, flowers red, 27 May 1938, Yu Dejun (T.T. Yü) 16021 (holotype in PE).

21. **Camellia hongkongensis** Seem., Trans. Linn. Soc. London 22: 342. 1859, *excl.* Gaudichaud 271.

Camellia japonica Champ. ex Benth. in Hook., Journ. Bot. 3: 309. 1851, *auct. non* L.

Thea hongkongensis (Seem.) Pierre, Fl. For. Cochinchine 2: sub. pl. 117. 1887.

Branchlets glabrous. Leaves oblong, 12 cm long, 3.8 cm wide. Flowers sessile, red; bracts and sepals 9-10, scarious, not completely deciduous; petals 6-7, 3.5 cm long; stamens with a filament tube, glabrous. Capsules globose, 3-3.5 cm in diameter, 3-locular, 1-2 seeds per locule, pericarp 3-4 mm thick.

Hong Kong: Chen Huanyong (W. Y. Chun) 9172; Huang Zhi (C. Wang) 30303;

1-3. *Camellia hongkongensis* Seem. 4-5. *Camellia japonica* L.
1. flowering branch; 2. perules and styles; 3. flower; 4. flowering branch; 5. pistil.

Chen Nianqu (N. K. Chun) 40030, 40280, 40286; Jiulong, Dalong Farm, Zhang Hongda (H. T. Chang) 6558, 6559.

Guangdong: Yangjiang, Hailing Island, Huang Zhi (C. Wang) 41713; Qingyuan, Zuo Jinglie (C. L. Tso) 22729.

Distribution: Offshore islands of Guangdong.

22. *Camellia cryptoneura* Chang, Acta Sci. Nat. Univ. Sunyatseni, monogr. ser. 1: 75. 1981.

Arbor circ. 8 m alta, ramulis glabris. Folia coriacea oblongo-oblanceolata vel oblonga 8-12 cm long 3-4 cm lata, apice acuminata basi cuneata decurrentia, supra in sicco olivaceo-viridia nitida, subtus flavobrunnea glabra, nervis lateralibus 6-jugis a costa sub angulo 35° abeuntibus, supra visibilibus subtus inconspicuis, margine superiore ¾ serrata, petiolis 1.5-2 cm longis glabris. Flores 1-2 terminales coccinei sessiles; bracteis et sepalis 8, exterioribus semiorbiculatis 3-6 mm longis sericeis, interioribus suborbicularibus 1.7 cm longis utrinque sericeis; petalis 8, exterioribus 2-3 pubescentibus, interioribus obovatis glabris basi connatis; staminibus 4-seriatis glabris; ovariis pilosis, stylis sparse puberulis apice 3-fidis. Capsula globosa 4 cm in diametro 3-locularis, 3-valvata dehiscens, valvis 4 mm (in sicco) crassis; semina 1-2 in quoque loculo.

Tree 8 m tall, branchlets glabrous; bud bracts ca. 15, puberulent. Leaves coriaceous, oblong-oblanceolate or oblong, 8-12 cm long, 3-4 cm wide, apices acuminate, bases cuneate, decurrent; olive-green above in the dry state, shiny; light brown below, glabrous; new leaves dark, glabrous; lateral veins 6 pairs, forming an angle of 35° from the midvein, visible on the upper surface, obscure on the lower surface; except for the basal ¼ the margins are sharply serrate, teeth separation 2 mm; petioles 1.5-2 cm long, glabrous. Flowers 1-2 terminal, pink, sessile; flower buds elliptic-ovate, 2.2 cm long, 1.3 cm wide; bracts and sepals 8, coriaceous; lower 3 perules semiorbicular, 3-6 mm long, outer surface pubescent; inner 5 perules suborbicular, 1.7 cm long, both surfaces pubescent; petals 8, back of outer 2-3 pubescent, inner 5-6 glabrous, obovate, basally connate; stamens in 4 series, outer whorl of stamens basally adnate with petals, inner whorl free, glabrous; ovaries pubescent; styles basally pubescent, apex deeply 3-cleft. Capsules globose, 4 cm in diameter, greyish-brown pubescent, 3-locular, pericarp 4 mm thick, 1-2 seeds per locule.

Guangxi: Damiao Shan, Miaozu Autonomous Prefecture, Rengan, Angui Village, Jinman Hamlet, Yuanbao Shan, elev. 1650 m, tree 8 m high, flower buds pink, 16 November 1938, He Xianzhang 82517; same loc., Lu Qinghua 3306; Longsheng, Huaping, Hekou, Chen Zhaozhou 51088; Guanyang, Shibajing, Chen Zhaozhou 52142; Damiao Shan, Chen Shaoqing (S. C. Chun) 16920; Longsheng, Zhong Jixin 91093 (holotype in SCBI); Lingui, Huang Shaoxiang and Chen Zhaozhou 50898.

23. *Camellia oviformis* Chang, Acta Sci. Nat. Univ. Sunyatseni, monogr. ser. 1: 75. 1981.

A *C. polyodonta* foliis oblongis, bracteis sepalisque minoribus, filamentis glabris, pericarpio suberoso differt.

Arbor parva circ. 8 m alta, cortice brunneo, ramulis glabris. Folia coriacea oblonga 11-16 cm longa 4-5.3 cm lata, apice abrupte acuta basi late cuneata, supra nitidula subtus glabris, nervis lateralibus 7-8-jugis impressis, margine acriter serrata, petiolis circ. 1 cm longis. Flores terminales rubri sessiles; bracteis sepalisque 10, extus sericeis, exterioribus 3 semilunatis 2-6 mm longis 4-11 mm latis, interioribus 7 orbicularibus vel ovoideis 8-14 mm longis; petalis 8 obovatis extus sericeis basi connatis; staminibus 4-5-seriatis, extimis dimidiis inferioribus connatis glabris; ovariis pilosis, stylis glabris apice 3-fidis. Capsula ovoidea 9 cm longa 7.5 cm lata, basi rotundata, apice attenuata obtusa brunnea 3-locularis, 3-valvata dehiscens, valvis suberosis 1-1.5 cm crassis; semina 1-2 in quoque loculo, 2.5-3.5 cm longa.

Tree 8 m tall, bark brown, branchlets glabrous. Leaves coriaceous, oblong, 11-16 cm long, 4-5.3 cm wide, apices acute, bases broadly cuneate; dark green above, slightly

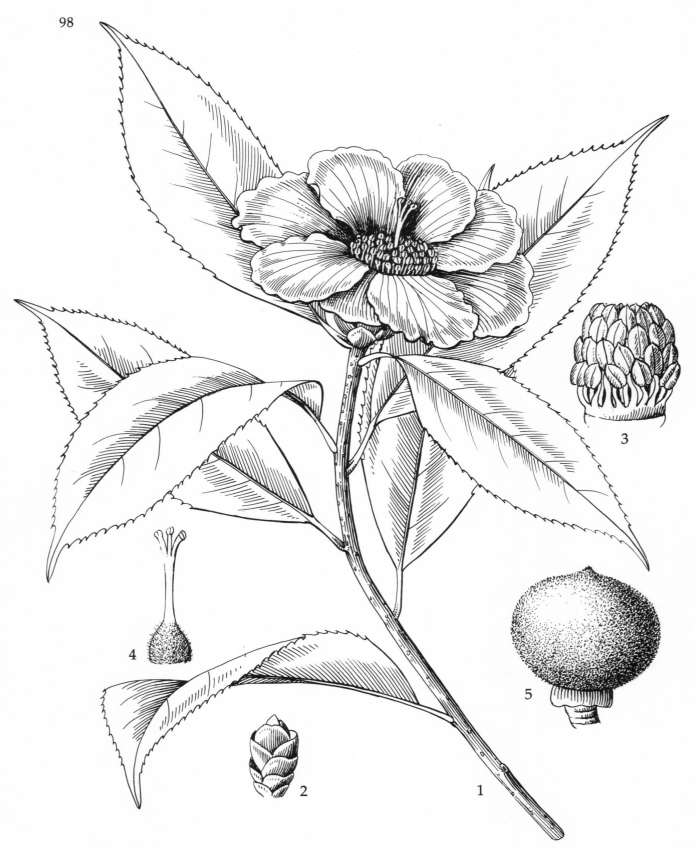

31 *Camellia cryptoneura* Chang
1. flowering branch; 2. flower bud; 3. stamens; 4. pistil; 5. capsule.

shiny; light green below, glabrous; lateral veins 7-8 pairs, sunken above; margins sharply serrate; petioles ca. 1 cm long, glabrous. Flowers terminal, red, sessile; bracts and sepals 10, exterior sericeous; 3 outer perules crescent shape, 2-6 mm long, 4-11 mm wide; 7 inner perules orbiculate to ovate, 8-14 mm long; petals 8, obovate, exterior sericeous, basally connate; stamens in 4-5 series; lower half of outer filament whorl connate into a short tube, glabrous; upper free half of filaments glabrous, ovaries pubescent; styles glabrous, apex deeply 3-cleft. Capsules ovate, 9 cm long, 7.5 cm wide, base rounded, apex slightly narrowly obtuse, brown, 3-locular, 1-2 seeds per locule, 3-valvate dehiscent; valves corky, 1-1.5 cm thick; seeds 2.5-3.5 cm long.

Guangxi: Lipu, Limu, Sujia Team, 28 October 1975, Liao Peilai 7501028 (holotype in SYS).

24. *Camellia compressa* Chang & Wen in Chang, Acta Sci. Nat. Univ. Sunyatseni, monogr. ser. 1: 76. 1981.

24a. *Camellia compressa* Chang & Wen in Chang var. *compressa*

Frutex vel arbor parva, ramulis glabris. Folia coriacea oblonga vel anguste elliptica, 10-14 cm longa 4-6 cm lata, apice acuminata vel abrupte acuta, basi late cuneata vel obtusa, supra nitida, subtus glabra, nervis lateralibus 7-8-jugis supra leviter impressis subtus visibilibus, margine serrulata, petiolis 1-1.2 cm longis. Flores subterminales rubri vel rubellini subsessiles, perulis 8-9 intimis 1-1.6 cm longis coriaceis ad medium griseosericeis; petalis 8-9 obovato-cordatis 3.5-5.5 cm longis exterioribus plus minusve sericeis; staminibus 3-4-seriatis, 2-3.5 cm longis extimis basi connatis glabris; ovariis 3-locularibus tomentosis, stylis staminibus brevibus apice 3-fidis. Capsulae compresse globosae 5-9 cm latae 3.5-7.5 cm altae 3-valvatae dehiscens, valvis suberosis 8-15 mm crassis, seminibus 2-3 in quoque loculo.

Shrub or small tree, branchlets glabrous. Leaves coriaceous, oblong or narrowly elliptic, 10-14 cm long, 4-6 cm wide, apices acuminate or acute, bases broadly cuneate or obtuse, shiny above, glabrous below; lateral veins 7-8 pairs, slightly sunken in the dry state, visible below; reticulating veins conspicuous, margins serrulate, petioles 1-1.2 cm long. Flowers subterminal, red or light red, sessile; perules 8-9, inner ones 1-1.6 cm long, coriaceous, middle of outer surfaces grayish-white sericeous; petals 8-9, obovate-cordate, 3.5-5.5 cm long, outer petals somewhat sericeous; stamens in 3-4 series, 2-3.5 cm long, outer filament whorl connate into a short tube, glabrous; ovaries 3-locular, velutinus; styles shorter than stamens, apices 3-cleft. Capsules oblate, 5-9 cm wide, 3.5-7.5 cm high, 3-valvate dehiscent, valves corky, soft, 8-15 mm thick, 2-3 seeds per locule.

Hunan: Long Shan, Houzitu, February 1979, Wen Xuankai s.n. (SYS no. 144701), 800201; same loc., Zanguo Commune, Kezhai Brigade, Dazao, September 1979, Wen Xuankai s.n. (SYS no. 147021); Baojing Xian, Boji Commune, Pijiang Brigade, Wen Xuankai 800305 (holotype in SYS).

This species is related to *C. phellocapsa* of eastern Hunan, but the leaves of the latter are elliptic, only upper half of margins serrrate, entire surface of perules pubescent, stamens in 5-6 series, capsules ovate to pyriform, seeds 4-5 per locule.

24b. *Camellia compressa* var. *variabilis* Chang & Wen in Wen, Acta Phytotax. Sin. 20: 227. 1982.

A typo foliis angastioribus, 10-16 cm longis, 3-5 cm latis, subtus in costa pilosis, floribus minoribus, ovariis 3-6-locularibus, stylis 3-6-fidis omnino pilosis differt.

This variety differs from the species by leaves long and narrow, 10-16 cm long, 3-5 cm wide, midvein pubescent below. Flowers smaller, ovaries 3-6 locular; styles 3-6-cleft, completely pubescent.

Hunan: Long Shan, elev. 800-1100 m, Wen Xuankai 800211 (holotype in SYS).

32 *Camellia compressa* Chang & Wen
1. flowering branch; 2. stamens; 3. pistil; 4. capsule.

Camellia setiperulata Chang & Lee
1. flowering branch; 2. pistil.

25. *Camellia setiperulata* Chang & Lee in Chang, Acta Sci. Nat. Sunyatseni, monogr. ser. 1: 77. 1981.

Species C. cryptoneurae affinis, qua foliis oblongo-oblanceolatis, bracteis et sepalis sericeis, non setosis, capsulis minoribus, pericarpio tenui differt.

Ramuli glabri. Folia oblonga 9-12 cm long 2.8-4 cm lata nitida, apice acuta basi cuneata, nervis lateralibus 6-8-jugis, margine serrulata, petioli 1-1.5 cm longi. Flores rubelli, bracteis sepalisque 9, intimis 2.5 cm longis fulvo-setosis coriaceis; petalis 6-7, extus setosis 5 cm longis, basi connatis; staminibus circ. 3 cm longis extimis basi connatis glabris; ovario setoso, stylo 3-fido; capsulis subglobosis 6-7 cm diam. pericarpio circ. 1 cm crasso.

Shrub or small tree, young branchlets glabrous. Leaves coriaceous, oblong, 9-12 cm long, 2.8-4 cm wide, both ends acute, both surfaces shiny, lateral veins 6-8 pairs, margins serrulate, petioles 1-1.5 cm long. Flowers light pink; bracts and sepals 9, coriaceous, to 2.5 cm long, densely yellowish-brown pubescent; petals 6-7, 5 cm long densely pubescent; stamens 3 cm long, glabrous; ovaries densely pubescent, styles 3-cleft. Capsules subglobose, 6-7 cm in diameter, pericarp ca. 1 cm thick.

Hunan: Ningyuan, Jiuni Shan, elev. 1100 m, wet area by a mountain stream, Xiao Zuncun 65001 (holotype in HNTC).

Similar to *C. cryptoneura*, but the latter has leaves oblong to oblanceolate, bracts and sepals pubescent, back of petals glabrous, capsules smaller, pericarp thinner.

26. *Camellia saluenensis* Stapf ex Bean, Trees and Shrubs, ed. 3, 66. 1933.
 Thea camellia var. *lucidissima* Lév., Cat. Pl. Yunnan, 27. 1917.
 Camellia speciosa Hort., Gard. Chron., ser. 3, 87: 297. 1930.
 Thea pitardii (Cohen-Stuart) Rehd.; Hand.-Mazz., Symb. Sinicae 7: 393. 1931, *pro parte*.
 Thea pitardii var. *lucidissima* (Lév.) Rehd., Journ. Arn. Arb. 15: 99. 1934.
 Camellia pitardii var. *lucidissima* (Lév.) Rehd., Journ. Arn. Arb. 18: 223. 1939.
 Camellia saluenensis f. *minor* Sealy, Rev. Gen. Camellia, 185. 1958.

Branchlets pubescent. Leaves narrowly oblong, 1.5-2.5 cm wide, both ends attenuate, apices not sharply acute, margins not sharply serrated. Capsules small, tricoccus, with three shallow grooves, some bicoccus or unicoccus; pericarp thin, less than 2 mm thick.

Yunnan: Zhaotong, Li Xiwen 67; Songming, Yu Pinghua 117; between Shuangbai and Guangtong, Yin Wenqing 861; Dali, Wang Hanchen (H. C. Wang) 4132, 4172, 4376, 4582; western Yunnan, G. Forrest 8640; Chengjiang, Chen Huanyong (W. Y. Chun) 1153; Liu Shenno (S. N. Liu) 15745, 15748, 16225, 23119; Songming, Xin Jingsan (G. S. Sin) 51346; Songming, Qiu Bingyun 51689.

27. *Camellia boreali-yunnanica* Chang, Acta Sci. Nat. Univ. Sunyatseni, monogr. ser. 1: 78. 1981.

A *C. pitardii* et var. *yunnanica* differt foliis crassioribus, nervis obscuris, floribus minoribus, petalis 9-10 angustioribus 3 cm longis, 1-1.2 cm latis leviter connatis tubis coroll arum 2-3 mm longis, staminibus 1.4 cm longis, capsulis oblatis.

Arbor, ramulis glabris. Folia crasse coriacea lanceolata 6-9 cm longa 2-2.5 cm lata, apice caudato-acuminata basi obtusa, supra nitida basi glabra, nervis lateralibus 7-8-jugis utrinque inconspicuis, margine serrulata, petiolis 5-8 mm longis. Flores rubri terminales, pedicellis 3-4 mm longis; bracteis sepalisque 6-10, maximis 1.2 cm longis; petalis 9-10 obovatis 3 cm longis 1-1.2 cm latis, basi 2-3 mm connatis; staminibus 1.4 cm longis, tubo filamentorum 1 cm longio glabro; ovariis pilosis, stylis 1.2 cm longis apice 3-fidis. Capsula oblata 2.5-3 cm diam., 3-valvata dehiscens, valvis 4-6 mm crassis; semina 1-2 in quoque loculo.

Small tree 2-4 m tall, branchlets glabrous. Leaves thickly coriaceous, lanceolate, 6-9 cm long, 2-2.5 cm wide, apices caudate-acuminate, bases obtuse, shiny above, yellowish-brown below; lateral veins 7-8 pairs, obscure on both surfaces; margins

1-2. *Camellia saluenensis* Stapf ex Bean 3-4. *Camellia hiemalis* Nakai
1. flowering branch; 2. flower; 3. flowering branch; 4. perules and pistil.

serrulate, petioles 5-8 mm long. Flowers red, 3 cm long; pedicels 3-4 mm long, covered by 6-10 bracts and sepals; petals obovate, 9-10, 3 cm long, 1-1.2 cm wide, basally connate for 2-3 mm; stamens 14 mm long, filament tube 1 cm long, glabrous; ovaries pubescent; styles 1.2 cm long, apically 3-cleft. Capsules oblate, 2.5-3 cm in diameter, 3-valvate dehiscent, valves 4-6 mm thick, 1-2 seeds per locule.

Yunnan: Heqing, Songgui, Maer Shan, 21 April 1939, Feng Guomei (K. M. Feng) 761 (holotype KUN); Yanjin, Chengfeng Commune, KUN Northeast Yunnan Group 1095.

Sichuan: Huili, Chuan Jingxi 855.

SUBSECT. II.
Lucidissima Chang, Acta Sci. Nat. Univ. Sunyatseni, monogr. ser. 1: 79. 1981.

Ovaria glabra, pericarpio plano glabro lignoso, staminibus glabris.
Ovaries glabrous; pericarp smooth, somewhat woody, glabrous; stamens glabrous.
6 species distributed in China and Japan.
Type: *C. lucidissima* Chang

28. *Camellia chekiangoleosa* Hu, Acta Phytotax. Sin. 10: 131. 1965.
Bracts and sepals 14-16, petals 7, ovaries glabrous. Fruit oblate, 3-5-locular, 3-8 seeds per locule.

Zhejiang: Kaihua, Wang Jingxiang 0001 (holotype in PE); Lishui, Wang Jingxiang s.n.; Lishui, Feng Jinyong 1054; Hangzhou, Zhang Hongda (H. T. Chang) 5165.

Jiangxi: Qiannan, Liu Xinqi (S. K. Lau) 4389.

Fujian: Chongan, Wang Mingin 3381; Tang Ruiyong 489; Fuding, Yang Cifu 703.
Hunan: Qi Zhongjing 0025.

This species is very close to *C. japonica* except the latter has only 10 bracts and sepals, petals 5-6, seeds smaller.

29. *Camellia lucidissima* Chang, Acta Sci. Nat. Univ. Sunyatseni, monogr. ser. 1: 79. 1981.
A *C. japonica* foliis nitidissimis majoribus, floribus majoribus petalis 5-6.5 cm longis, staminibus liberis differt; a *C. chekiangoleosa* bracteolis et sepalis paucioribus 9-10, petalis 6-7, staminibus liberis recedit.

Arbor, ramulis glabris. Folia coriacea elliptica 7-13 cm longa 3-5 cm lata, apice abrupte acuta basi late cuneata, supra lucidissima subtus glabra, nervis lateralibus 6-7-jugis, margine superiore serrulata, petiolis 1-1.5 cm longis. Flores terminales rubri sessiles; bracteis et sepalis 9-10 exterioribus 4 minoribus 5-7 mm longis, interioribus ovoidea ad orbiculatis 2-2.5 cm longis sericeis; petalis 6-7, 5-6.5 cm longis; staminibus 2.5-3 cm longis, extimis subliberis glabris; ovariis glabris 3-locularibus, stylis apice 3-fidis. Capsula subglobosa vel leviter compressa 4-6 cm in diametro, pericarpio laevi 3-valvato dehiscens, valvis 5-7 mm crassis.

Small tree, branchlets glabrous. Leaves coriaceous, elliptic, 7-13 cm long, 3-5 cm wide, apices acute, bases broadly cuneate, extremely shiny above, glabrous below, lateral veins 6-7 pairs, upper half of margins serrulate, petioles 1-1.5 cm long. Flowers terminal, sessile, red; bracts and sepals 9-10, pubescent; outer 4 perules small, 5-7 mm long; inner 5-6 perules ovoid, 2-2.5 cm long; petals 6-7, 5-6.5 cm long; stamens 2.5-3 cm long, outer filament whorl free, glabrous; ovaries glabrous, 3-locular, styles apically 3-cleft. Capsules subglobose, 4-6 cm in diameter, pericarp shiny and smooth.

Jiangxi: Lijiang, Yang Xiangxue (C. X. Yang) s.n. (PE no. 940326); Nanfeng, Xanxi Commune, Junfeng Shan, elev. 900-1200 m, Yang Xiangxue 650455 (holotype in PE); HLG 3061, 5503.

Zhejiang: Qingyuan, Zhang Zhaofang and Huang Zhengzheng 315; Ruian, Zhang Shaoyao 6643; Taishun, Zhang Shaoyao 5566.

Camellia chekiangoleosa Hu
1. flowering branch; 2. stamens; 3. anthers; 4. pistil; 5. capsule; 6. seeds.

1. *Camellia lucidissima* Chang
2. *Camellia subintegra* P. C. Huang
1. flowering branch; 2. fruiting branch.

30. ***Camellia magnocarpa*** (Hu & Huang) Chang, Acta Sci. Nat. Univ. Sunyatseni, monogr. ser. 1: 81. 1981.

Camellia semiserrata var. *magnocarpa* Hu & T. C. Huang in Hu, Acta Phytotax. Sin. 10: 137. 1965.

A *C. semiserrata* differt bracteis et sepalis glabris, ovariis glabris, fructibus majoribus, seminibus 5 in quoque loculo.

Arbor circ. 8.5 m alta, truncus 50 cm in diametro, ramulis glabri. Folia coriacea elliptica vel obovato-elliptica 9-22 cm longa 4-8 cm lata, apice abrupte acuminata basi late cuneata vel subrotundata, supra nitida subtus glabra, nervis lateralibus 6-7-jugis, margine superiore serrata, petiolis 1.5-2.5 cm longis. Flores terminales rubri sessiles; bracteis et sepalis glabris; petalis 8-9 late obovatis 3 cm longis 4.5 cm latis; staminibus 5-seriatis, extimis inferioribus connatis sparse pubescentibus; ovariis glabris, stylis 2.5 cm longis, apice 3-fidis. Capsula globosa 6-12 cm in diametro 3-locularis, 3-valvata dehiscens, valvis 1.3-1.7 cm crassis, columella rubusta; semina 2-5 in quoque loculo, semiorbiculata 1.2 cm longa.

Small tree ca. 8.5 m tall, branchlets glabrous; buds long-ovate, completely glabrous. Leaves thickly coriaceous, elliptic, 9-22 cm long, 4.8 cm wide, apices acuminate with a 1.5-2 cm long cauda, bases broadly cuneate or suborbicular; dark green above, shiny in the dry state; light green below, glabrous; lateral veins 6-7 pairs, slightly raised on both surfaces in the dry state; margins sparsely sharply serrate, teeth separation 2-6 mm, teeth 1-1.5 mm long; petioles 1.5-2.5 cm long, upper surface canaliculate, glabrous. Sepals orbicular, outer surfaces glabrous, deciduous; petals 8-9, 3 cm long, 4.5 cm wide; stamens puberulent, ovaries glabrous, styles apically 3-cleft. Capsules globose, sessile, 3-locular, 2-5 seeds per locule, 3-valvate dehiscent, pericarp 1.3-1.7 cm thick, columella thick; seeds semiglobose, 1.2 cm long, light brown.

Guangdong: Ruyuan, Tianjing Shan, on the way to the T.V. tower, elev. 1400 m, under forest, August 1975, Zhang Hongda (H. T. Chang) 6030, 6046; Yang Shan, Zhaotianshi, Deng Liang 1192, 1468; Guangning, Mou Ruhuai 40024.

Guangxi: Cangwu, Huang Zuojie (T. C. Huang) 2088 (holotype in SCBI); Shangsi, Zhang Zhaoqian 13913; Teng Xian, Feng Jinyong 1014.

31. ***Camellia japonica*** L., Sp. Pl. 2: 698. 1753.
Thea camellia Hoffm., Verz. Pflz., 117. 1824.
Thea japonica (L.) Baill, Hist. Pl. 4: 229. 1873.
Thea hozanensis Hay., Icon. Pl. Formos. 7: 2. 1918.
Camellia hozanensis (Hay.) Hay., Icon. Pl. Formos. 8: 11. 1919.
Camellia japonica var. *hozanensis* (Hay.) Yamamoto, Sylvia 5: 35. 1934.
For additional synonyms see Sealy, Rev. Gen. Camellia, 175. 1958.

31a. ***Camellia japonica*** L. ssp. *japonica* var. *japonica*

Leaves elliptic, 7-10 cm long; lateral veins 6-8 pairs, often slightly raised. Perules 9-10; petals 6-7 in wild form, light pink, cultivated forms with many petals, red or white; filaments in a short tube, ovaries glabrous. Capsules 3-locular, 3-4 cm in diameter.

Sichuan: Emei Shan, Yang Guanghui 53859; same loc., SZ s.n. (SCBI no. 269346) (These two specimens are from wild plants).

Shandong: Lushan, Jiao Qiyuan 2855 (wild).

Jiangxi: Lichuan, Huixianfeng, Wang Mingjin 2154 (wild).

Guangdong: Guangzhou, Zhang Hongda (H. T. Chang) 6021; McClure 20846 (cultivated).

Guangxi: Guilin, McClure 20608 (cultivated).

Fujian: Tang Ruiyong s.n. (SYS no. 134714) (cultivated).

Japan: Ryukyu Islands, Nagomachi Ins. Okinawa, T. Kanashiro s.n. (SYS no. 113169).

The Chinese specimens from Sichuan, Shandong and Jiangxi listed above are wild species; they have single petals and perfectly formed filament tubes and stamens.

31b. *Camellia japonica* ssp. *japonica* var. *macrocarpa* Masamune, Trans. Nat. Hist. Soc. Formosa 23: 205. 1933.

Camellia japonica var. *spontanea* Masamune, Prelim. Rep. Veg. Yakusima, 96. 1929.
Camellia hayaoi Yanajida, Journ. Jap. Forest. Soc. 15: 132. 1933.

Leaves relatively small, 5.5-7.8 cm long, 2-3.5 cm wide. Capsule 5-7 cm in diameter, pericarp 1.5-2 cm thick.

Japan: Hondo, Sagami Pref., Kamakura, K. Hisauti s.n. (SYS no. 95114).

31c. *Camellia japonica* ssp. *rusticana* (Honda) Kitamura, Acta Phytotax. et Geobot. Kyoto 14: 61. 1950.

Camellia rusticana Honda, Biosphaera 1: 97. 1947.

This subspecies is pubescent and the petioles are relatively short, flower perules relatively short, petals connate and relatively tiny, stamens relatively short.

Distribution: Japan.

32. *Camellia subintegra* P. C. Huang in Chang, Acta Sci. Nat. Univ. Sunyatseni, monogr. ser. 1: 83. 1981.

A *C. japonica* differt foliis angustioribus subintegris vel pauloserrulatis, capsulis ellipsoideis; a *C. chekiangoleosa* Hu foliis angustioribus subintegris, perulis et petalis minoribus, seminibus 1-3 in quoque loculo differt.

Arbor parva, ramulis glabris. Folia coriacea anguste oblonga vel lanceolata, 8-11 cm longa 2-3.5 cm lata, apice acuminata basi cuneata, supra nitida subtus glabra, nervis lateralibus 6-7-jugis, margine subintegra vel superiore ⅓ remote serrulata, petiolis 1.5 cm longis. Flores 1-2 terminales rubri 8-9 cm in diametro sessiles; bracteis et sepalis 10-12 coriaceis 1.2-2 cm longis sericeis; petalis 5-6, circ. 3.5-5.7 cm longis basi connatis; staminibus 4-5-seriatis, extimis inferioribus connatis glabris; ovariis glabris 3-4-locularibus, stylis 2 cm longis, apice 3-fidis. Capsula ovoidea 3-6 cm longa 3-4-valvata dehiscens, valvis 1-2 cm crassis; semina 1-3 in quoque locula.

Small tree, branchlets glabrous. Leaves coriaceous, narrowly oblong or lanceolate, 8-11 cm long, 2-3.5 cm wide, apices acuminate, bases cuneate, shiny above, glabrous below, lateral veins 6-7, margins subentire or upper ⅓ sparsely serrulate, petioles ca. 1.5 cm long. Flowers 1-2 terminal, red, 8-9 cm in diameter, sessile; bracts and sepals 10-12, coriaceous, 1.2-2 cm long, sericeous; petals 5-6, 3.5-5.7 cm long, basally connate; stamens in 4-5 series, outer filament whorl connate into a tube, free filaments glabrous; ovaries glabrous, 3-4 locular; styles 2 cm long, apically 3-cleft. Capsules ovoid, 3-6 cm long, 3-4-valvate dehiscent, valves 1-2 cm thick, 1-3 seeds per locule.

Jiangxi: Yichun, Anfu, Wang Mingxiu and Huang Pengcheng 23, 226 (holotype in NJTFC); Wugong Shan, Nanping, Inst. Bot. Jiangxi Team 1515, 1150.

33. *Camellia longicaudata* Chang & Liang in Chang, Acta Nat. Sci. Univ. Sunyatseni, monogr. ser. 1: 83. 1981.

Species *C. subintegrae* affinis, a qua differt foliis anguste lanceolatis longe caudatis, margine serrulatis, floribus minoribus.

Arbor parva, ramulis glabris. Folia coriacea anguste lanceolata 10-14 cm longa 2-3 cm lata, apice caudata, caudo 2-2.5 cm longo, basi cuneata interdum obtusa vel subrotundata, supra nitidula subtus glabra, nervis lateralibus 7-8-jugis ut costis supra impressis, margine dense serrulata, dentis inter se 1.5 mm remotis, petiolis 8-12 mm longis glabris. Flores terminales rubri sessiles; bracteis et sepalis 10, exterioribus 2 late ovatis puberulis 4-7 mm longis 4-9 mm latis, interioribus subrotundatis 2-2.8 cm longis sericeis; petalis 9 basi connatis; staminibus glabris basi connatis, tubo filamentorum brevi; ovariis glabris, stylis 1 cm longis, apice 3-fidis.

Small tree 8 m tall, branchlets glabrous; leaf buds long-ovate, outer bracts glabrous, inner bracts externally pubescent. Leaves coriaceous, narrowly lanceolate, 10-14 cm long, 2-3 cm wide, apices caudate with a 2-2.5 cm long cauda, bases broadly cuneate or subrotund; dark brown above in the dry state, not shiny and smooth, glabrous; yellowish-brown below, glabrous; lateral veins 7-8 pairs, midveins sunken above and protruding below; margins densely serrulate, teeth every 1.5 mm; petioles 8-12 mm long, glabrous. Flowers terminal, red, sessile; bracts and sepals 10; outer 2 perules broadly ovate, pubescent, 4-7 mm long, 4-9 mm wide; inner perules suborbicular, 2-2.8 mm long, exterior sericeous; petals 9, basally connate; stamens glabrous, forming a filament tube; ovaries glabrous; styles 1 cm long, glabrous, apically 3-cleft. Flowering period August to September.

Guangdong: Dinghu Shan, Longchuankeng, 3 August 1963, Ding Guanqi and Shi Guoliang 830 (holotype in SYS); Zhanjiang District, Nan Zhidi 2897.

Guangxi: Yao Shan, Guchen, Li Binggui 015.

Chapter 5

SUBGENUS *Thea*

SUBGEN. III.
Thea (L.) Chang, Acta Sci. Nat. Univ. Sunyatseni, monogr. ser. 1: 86. 1981.
  Thea L., Sp. Pl. 1: 515. 1753.

Floribus 1-3 axillaribus vel terminalibus mediis vel minoribus, pedicellatis; bracteis sepalisque distinctis, bracteolis 2-6 vel pluribus persistentibus vel caducis; sepalis 5-6 persistentibus; petalis 5-12 liberis vel leviter connatis; ovario 3-5 locularii; stylis 3-5 vel pluribus; capsulis 3-5-locularibus columnaris.

Flowers 1-3 axillary, pedicellate, medium to relatively small, bracts and sepals usually differentiated; bracts 2-6 or more, persistent or deciduous; sepals 5-6, persistent; petals slightly connate or free; stamens in 2-3 series, free or outer whorl connate; ovaries 3-5-locular, occasionally more; styles 3-5-parted or 3-5-cleft, sometimes 7-cleft. Capsules 3-5-locular, with a columella.

8 sections and 60 species.

Type: *C. sinensis* (L.) O. Kuntze

SECTION XI.
Corallina Sealy, Rev. Gen. Camellia, 132. 1958; Chang, Acta Sci. Nat. Univ. Sunyatseni, monogr. ser. 1: 86. 1981, *diagnosis emend.*

Flores 2-4 cm diam., bracteae et sepala indistincta 6-10 persistentia vel interdum 4, petala 5-10 connata, stamina 2-3-seriata, sublibera, ovaria glabra vel villosa 3-locularia stylis 3 liberis, capsula 3-locularis columnifera.

Flowers medium sized, 2-4 cm in diameter, white, axillary or terminal, sessile or short pedicellate; bracts and sepals not completely distinct, gradually grading from one to the other, persistent, 6-10, sometimes only 4; petals 5-10, connate; stamens in 2-3 series, 1 cm long, connate or the outer ones adnate with petals, the remaining free; ovaries glabrous or pubescent, 3-locular; styles 3-parted, free, rarely connate. Capsules 3-locular, with a columella.

11 species; 8 in southern China, the remaining 3 in central and southern Indochina.

Type: *C. corallina* (Gagnep.) Sealy

The main characteristics of this section are the bracts and sepals not completely differentiated, persistent, stamens nearly free; capsules 3-locular, with a columella; styles mostly free. This section shares characteristics with section *Chrysantha*, and perhaps both are in the same stage of parallel development. The flowers of *C. nitidissima* and *C. tonkinensis* have not been seen, so the characteristics of the bracts and sepals have been used to place them in this section.

KEY TO SECT. CORALLINA

1. Ovaries pubescent.
    2. Leaves petiolate, fruit tricoccus, pericarp thin, seeds globose.
        3. Flowers pedicellate; leaves greater than 10 cm long, back somewhat pubescent; fruit flattened, 3 cm wide.
            4. Branchlets pubescent, bracts and sepals 4, flowers light red, leaves oblong.................................... 1. *C. corallina* (Gagnep.) Sealy
            4. Branchlets glabrous, bracts and sepals 6, flowers white; leaves elliptic, chartaceous.................... 2. *C. tonkinensis* (Pitard) Cohen-Stuart
        3. Flowers sessile; leaves 6-9 cm long, oblong, glabrous; fruit globose, 2 cm in diameter.. 3. *C. wardii* Kobuski
    2. Leaves subsessile, bases cordate or auriculate, 3-5 cm long; fruit globose, seeds pubescent... 4. *C. pilosperma* Liang
1. Ovaries glabrous.
    5. Leaves oblong, elliptic or obovate, bases cuneate; branchlets glabrous, sepals coriaceous.
        6. Leaf margins serrate.
            7. Leaves oblong or elliptic, apices acute, back surfaces not atropunctate.
                8. Leaves oblong, 2.5-3.7 cm wide; capsules furfuraceous, pedicels extremely short.................... 5. *C. fleuryi* (A. Chev.) Sealy
                8. Leaves elliptic, 3-5 cm wide; capsules smooth, pedicels conspicuous .... 6. *C. nitidissima* Chi
            7. Leaves obovate, apices rounded, hard coriaceous, back of leaves atropunctate .... 7. *C. paucipunctata* (Merr. & Chun) Chun
        6. Leaves entire, oblong, lateral veins 4-6 pairs .... 8. *C. lienshanensis* Chang
    5. Leaves long-ovate, bases subrotundate or broadly cuneate; branchlets pubescent, sepals membranous.
        9. Leaves 4-7 cm long, lateral veins 5-6 pairs; petals 5-9.
            10. Petals 5, 1 cm long; leaves 4-5.5 cm long...... 9. *C. pentamera* Chang
            10. Petals 9, 2 cm long; leaves 5-7 cm long ...... 10. *C. scariosisepala* Chang
        9. Leaves 5-7 cm long, lateral veins 9-11 pairs; petals 6-8. .... 11. *C. acutiserrata* Chang

1. ***Camellia corallina*** (Gagnep.) Sealy, Rev. Gen. Camellia, 132. 1958.
    Thea corallina Gagnep., Not. Syst. 10: 126. 1942.
    Branchlets pubescent. Leaves oblong, 7-12 cm long, 2.5-4 cm wide, midrib of back surface pubescent. Flowers born towards the branch terminal in leaf axils; pedicels 7 mm long, lacking bracts; sepals 4, paired; petals 9-10, red, 1 cm long; stamens free, ovaries pubescent, styles 3-parted.
    Distribution: Vietnam.

2. ***Camellia tonkinensis*** (Pitard) Cohen-Stuart, Meded. Proefst. Thee 40: 67. 1916.
    Thea tonkinensis Pitard in Lecomte, Fl. Gén. Indo-Chine 1: 343. 1910.
    Branchlets glabrous. Leaves elliptic, 9-13 cm long, 2.5-5 cm wide, thinly chartaceous, puberulent along midrib of lower surface, pedicels 7 mm long. Bracts and sepals 6, ovaries pubescent; styles 3-parted, free. Capsules tricoccus, flattened, 3.8 cm wide.
    Distribution: Vietnam.

3. ***Camellia wardii*** Kobuski, Brittonia 4: 114. 1941.
    Branchlets glabrous. Leaves orbicular, 6-9 cm long, apices caudate-acuminate, bases cuneate, margins sharply serrate. Flowers sessile; bracts and sepals 8-9, persistent; petals 6-8, 2 cm long, basally connate; stamens 1.2 cm long, nearly free, ovaries pubescent; styles 3-parted, 6-8 mm long. Capsules tricoccus, 2 cm in diameter, pericarp thin.

Sichuan: Tianquan, Songjia Village, Hu Wenguang and He Zhu 12021.

Yunnan: Jingdong, Fengguan Shan, Li Minggang 945; Yingjiang Xian, Xima Commune, Nabangba, Tao Guoda 13131.

Distribution: Yunnan, Sichuan, also reaching the northern part of Burma.

4. *Camellia pilosperma* Liang, Acta Phytotax. Sin. 17 (4): 95. 1979.

Frutex, ramulis hirsutis, foliis ovato-oblongis 2-5.5 cm longis basi auriculatis amplexcaulibus, bracteolis et sepalis 7-8 persistentibus glabris petalis 5-7 subliberis 1.9-2.2 cm longis, staminibus 2-3-seriatis extimis in triente infero connatis, ovario hirstello 3-loculari, stylis 3 subliberis 1-1.3 cm longis, capsulis minoribus subglobosis vel depresse deltoideo-globosis 7-12 mm in diametro, seminibus solitariis vel rarius bini in quoqueloculo puberulis.

Shrub 1-3 m tall; branchlets brown hirsute, old branches gray, with exfoliating stringy bark. Leaves coriaceous, elliptic or ovate-oblong, 2-5.5 cm long, 1.5-2.5 cm wide, apices subacute or slightly obtuse; bases auriculate, auricles 3-5 mm long; dark brown above in the dry state, not shiny, base of the midvein pubescent; brown below, midvein pilose; lateral veins 7-8, sunken above, slightly protruding below; margins sparsely serrate, teeth separation 2.5-3.5 mm; petioles extremely short. Flowers 1-2 terminal or axillary, pedicels extremely short; perules 7-8, outer perules 1.5-2 cm long, semi-orbicular or broadly ovate, glabrous; inner perules broadly ovate or ovoid, 3.5-4 mm long, glabrous, bases nearly free, persistent; petals 5-7, obovate, 1.9-2.2 cm long, nearly free, apices emarginate, glabrous, white; stamens 8-10 mm long, outer filament whorl basally connate for 2 mm, glabrous; inner whorls completely free; anthers basifixed, yellow; ovaries pubescent, 3-locular; styles 3-parted, 1-1.3 mm long, completely free. Capsules globose, 7-12 mm in diameter, 1-locular, 1 seeded, usually indehiscent, pericarp thin, surface puberulent.

Guangxi: Zhaoping, Nanyong Village, elev. 100-500 m, Liang Shengye (S. Y. Liang) 6505259 (holotype in GXFI; Isotypes in GXMI, PE, SYS); Zhaoping, Danao Shan Forest Center, November 1958, Li Yinkun 402349, 402425.

A number of characters of this species differ from what is normal for this section including the anthers basifixed, seeds pubescent and leaf bases auriculate.

5. *Camellia fleuryi* (A. Chev.) Sealy, Kew Bull. 1949 (2): 217. 1949.
Thea fleuryi A. Chev., Bull. Econ. Indochine 21: 531. 1919.

Branchlets glabrous. Leaves orbicular, 7-11 cm long, lateral veins 6 pairs. Bracts and sepals 8-9; petals 5, 1-1.5 cm long; stamens free, ovaries glabrous; styles 3-parted, free. Capsules furfuraceous.

Distribution: Vietnam.

6. *Camellia nitidissimia* Chi, Sunyatsenia 7: 19. 1948.

Leaves elliptic, 11 cm long, 5 cm wide. There are no specimens of this species with flowers. Fruit oblate, 5 cm in diameter, becoming black in the dry state; pedicels thick, to almost 1 cm long, persistent bracts and sepals 11-12.

Guangxi: Shiwan Dashan, Zuo Jinglie (C. L. Tso) 23483 (holotype in SYS); Fangcheng, Jiang Shan, Taiping, Liang Kui 69451.

7. *Camellia paucipunctata* (Merr. & Chun) Chun, Sunyatsenia 4: 187. 1940.
Thea paucipunctata Merr. & Chun, Sunyatsenia 2: 285. 1935, *excl.* H. Fung 20336.

Small tree, branches glabrous. Leaves coriaceous, obovate-elliptic, 10 cm long, apices rounded or cuneate, back surfaces atropunctate. Pedicels extremely short; perules 8-9, not differentiated, persistent; petals 6, 2.5 cm long, white; stamens nearly free, ovaries glabrous; styles 3-parted, free. Capsules globose.

Hainan: Liu Xinqi (S. K. Lau) 5946; Ya Xian, Chen Nianqu (N. K. Chun) and Zuo Jinglie (C. L. Tso) 44625 (holotype in SYS, isotype in A).

Distribution: Hainan.

8. *Camellia lienshanensis* Chang, Acta Sci. Nat. Univ. Sunyatseni, monogr. ser. 1: 89. 1981.

A ceteris sectionis foliis integris, stylis connatis apice 3-fidis recedit.

Frutex circ. 3 m altus, ramulis glabris. Folia coriacea anguste elliptica 6-9.5 cm longa 2.5-3.5 cm lata, apice acuminata basi cuneata, supra olivaceo-viridia nitida subtus glabra, nervis lateralibus 4-6-jugis utrinque haud conspicuis, margine integra, petiolis 1.2-1.6 cm longis. Flores terminales sessiles; bracteis sepalisque 10, inferioribus 2-3 semilunatis 1.5-3 mm longis 4-7 mm latis, ceteris orbiculatis vel obovatis 7-12 mm longis 9-13 mm latis, utrinque puberulis; petalis 6 exterioribus 2 pubescentibus, ceteris glabris; staminibus basi ad petalum adnatis exceptis liberis; ovariis glabris, stylis connatis circ. 1 cm longis apice 3-fidis.

Shrub 3 m tall, branchlets glabrous, older branches grayish-white. Leaves coriaceous, narrowly elliptic, 6-9.5 cm long, 2.5-3.5 cm wide, apices acuminate, bases cuneate; olive-green above, shiny, glabrous; light green below, glabrous; lateral veins 4-6 pairs, equally obscure on both surfaces; entire; petioles 1.2-1.6 cm long, glabrous. Flowers terminal, sessile; bracts and sepals 10, lower 2-3 semilunate, 1.5-3 mm long, 4-7 mm wide; remaining 7-8 perules from semiorbicular to orbicular or obovate, 7-12 mm long, 9-13 mm wide, inner and outer surfaces puberulent; petals 6, not open, outer 2 pubescent, others glabrous; stamens glabrous; filaments basally slightly adnate with petals, remainder free; ovaries glabrous; styles ca. 1 cm long, apices shallowly 3-cleft. Flowering period October.

Guangdong: Lian Shan, Dalong Shan, October 1945, Chen Shaoqing (S. C. Chen) 5678 (holotype in SCBI); Liannan, Jinkeng, Huangdong, elev. 700 m, tree 7 m tall, 20 September 1958, Tan Peixiang 59565.

In section *Corallina* this is the only species with entire margins. A single style is also unusual for this section. This species is tentatively placed in this section because the fruit has not yet been seen.

9. *Camellia pentamera* Chang, Acta Sci. Nat. Univ. Sunyatseni, monogr. ser. 1: 90. 1981.

A ceteris sectionis bracteis 5, sepalis 5, petalis 5, staminibus 25, 1-seriatis distincta.

Arbor circ. 15 m alta, trunco 20 cm diam., ramulis pubescentibus. Folia coriacea ovato-elliptica 4-5.5 cm longa 1.7-2.3 cm lata, apice acuminata, acumene obtuso, basi obtusa vel subrotundata, supra opaca ad costam puberula, subtus glabra, nervis lateralibus 5-6-jugis inconspicuis margine serrata, petiolis 4-5 mm longis glabris. Flores soliterii terminales albi subsessiles; perulis 10, exterioribus 5 (bracteis) 2-4 mm longis glabris, interioribus 5 (sepalis) late ovatis 6-8 mm longis apice sparse puberulis margine scariosis; petalis 5 obovatis 1.7-2 cm longis basi connatis; staminibus 25, 1-seriatis, 1.2-1.5 cm longis, basi leviter connatis; ovariis glabris 3-locularibus, stylis 3 subliberis 1.5 cm longis, basi 5 mm connatis.

Tree 15 m tall, trunk diameter 20 cm, branchlets pubescent. Leaves coriaceous, ovate-elliptic, 4-5.5 cm long, 1.7-2.3 cm wide; apices acuminate, tip cuneate; bases cuneate or subrotundate; shiny above, midveins with remnants of pubescence above; glabrous below; lateral veins 5-6 pairs, obscure; margins serrate; petioles 4-5 mm long, pubescent. Flowers terminal, white, subsessile; bracts 5, triangular-ovate, 2-4 mm long, glabrous; sepals 5, broadly ovate, 6-8 mm long, extreme apex slightly pubescent, margins membranous; petals 5, obovate, 1.7-2 cm long, basally connate; stamens 25, 1 series, 1.2-1.5 cm long, bases slight connate; ovaries glabrous, 3-locular; styles 3-parted, 1.5 cm long, basally connate for 5 mm.

Yunnan: Wen Shan, Laojun Shan, hillside behind Yaodian Forest Farm, elev. 2450 m, on the slope of the mountain ridge, Feng Guomei (K. M. Feng) 22535 (holotype in KUN).

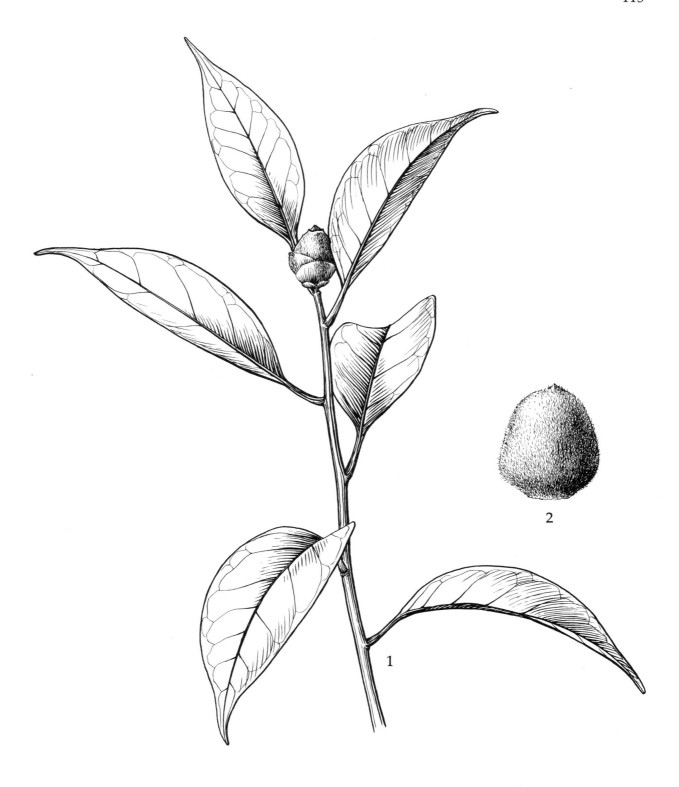

Camellia lienshanensis Chang
1. branch in bud; 2. sepal.

10. *Camellia scariosisepala* Chang, Acta Sci. Nat. Univ. Sunyatseni, monogr. ser. 1: 91. 1981.

A *C. pentamera* differt foliis tenuibus, petalis 9 majoribus 9 mm longis, stylis 4 perfecte liberis.

Arbor parva circ. 3-4 m alta, ramulis pubescentibus. Folia tenuia ovato-elliptica 5-7 cm longa 2-3 cm lata, apice acuminata vel caudata, basi late cuneata, supra opaca subtus primo ad costam pubescentia nox glabrescentia, nervis lateralibus circ. 6-jugis, margine acriter serrata, petiolis 4-6 mm longis pubescentibus. Flores albi subsessiles; bracteis et sepalis 9-10 suborbiculatis 4-9 mm longis utrinque glabris, margine scariosis; petalis 9 obovatis circ. 2 cm longis; staminibus 1-1.2 cm longis basi leviter connatis; ovariis glabris, stylis 4 liberis 1-1.2 cm longis.

Small tree 3-4 m tall, branchlets pubescent. Leaves thin, ovate-elliptic 5-7 cm long, 2-3 cm wide, apices acuminate or caudately acuminate, bases broadly cuneate, shiny above, midveins at first pubescent below, lateral veins ca. 6 pairs, margins sharply serrate; petioles 4-6 mm long, pubescent. Flowers white, subsessile; bracts and sepals 9-10, suboribuclar 4-9 mm long, both surfaces glabrous, margins thinly membranous; petals 9, obovate, ca. 2 cm long; stamens 1-1.2 cm long, slightly connate; ovaries glabrous; styles 4-parted, free, 1-1.2 cm long.

Yunnan: Yongping, Shayang, above Cheng Guosi, March 1958, Yunnan Forest Management Brigade, 4th Team 161 (holotype in KUN).

11. *Camellia acutiserrata* Chang, Acta Sci. Nat. Univ. Sunyatseni, monogr. ser. 1: 91. 1981.

A *C. pentamera* differt nervis lateralibus pluribus, petalis pluribus, stylis perfecte liberis; a *C. scariosisepala* Chang nervis lateralibus pluribus, petalis minoribus, staminibus brevioribus differt.

Arbor parva circ. 4-5 m alta, ramulis pubescentibus. Folia coriacea ovato-lanceolata 5-7 cm longa 2.5-3.3 cm lata, apice caudato-acuminata basi rotundata, supra atro-viridia opaca, subtus ad costam puberula, nervis lateralibus 9-11-jugis, margine acriter serrata, petiolis 5-7 mm longis puberulis. Flores axillares vel terminales albi, pedicellis brevissimis; bracteis 5 late ovatis 2 mm longis glabris; sepalis 5, suborbiculatis 4-11 mm longis scariosis glabris; petalis 6-8 circ. 2 cm longis glabris; staminibus brevibus basi leviter connatis; ovariis glabris 3-locularibus, stylis 3 liberis. Capsula globosa 2.5-3 cm in diametro 3-valvata dehiscens, valvis 5 mm crassis; semina solitaria in quoque loculo.

Small tree 4-5 m tall, branchlets pubescent. Leaves coriaceous, ovate-lanceolate, 5-7 cm long, 2.5-3.3 cm wide, apices caudate-acuminate, bases rounded, not shiny above, pubescent along the midvein below, lateral veins 9-11 pairs, margins sharply serrate, petioles 5-7 mm long. Flowers subsessile; bracts 5, broadly ovate, 2 mm long; sepals 5, suborbicular, 4-11 mm long, glabrous, scarious; petals 6-8, 2 cm long, white, glabrous; stamens short; ovaries glabrous, 3-locular; styles 3-parted. Capsules globose, 2.5-3 cm in diameter, 3-valvate dehiscent, valves 5 mm thick, 1 seed per locule.

Guizhou: Ceheng, Liujiadong, Cao Ziyu 456 (holotype in PE).

SECTION XII.
Brachyandra Chang, Acta Sci. Nat. Univ. Sunyatseni, monogr. ser. 1: 92. 1981.

A sect. *Corallina* differt staminibus multo brevioribus filamentis robustis, basi connatis, stylis brevioribus 1-4 mm longis.

Flowers small, terminal or axillary, sessile or petiolate; bracts and sepals 6-10, not distinctly separate, persistent; petals 5-10, basally connate; stamens in 2 series, 4-8 mm long; filament robust, basally connate; ovaries glabrous or pubescent, 3-5-locular; styles extremely short, 1-4 mm long, free or somewhat connate. Capsules 3-5-valvately dehiscent.

12 species; 9 in southern China and southwestern China, 3 in Indochina.
Type: *C. brachyandra* Chang
This section is very close to section *Corallina*, except the stamens are shorter, basally connate forming a short tube, filaments sometimes very thick; styles shorter, not more than 1-4 mm long.

KEY TO SECT. *BRACHYANDRA*

1. Ovaries pubescent.
    2. Branchlets pubescent; leaves oblong or long-ovate, 4-9 cm long; bracts and sepals 7-11.
        3. Leaves 6-9 cm long, lateral veins 7-8 pairs, bracts 3-4, petals 5, styles 1 mm long. . . . . 1. *C. muricatula* Chang
        3. Leaves 5-6 cm long, lateral veins 5-6 pairs, bracts 2, petals 7, styles 3-4 mm long . . . . 2. *C. szemaoensis* Chang
    2. Branchlets glabrous; leaves elliptic, 10 cm long; bracts and sepals 6-9; pericarp 2-7 mm thick.
        4. Bracts and sepals 6, stamens 6 mm long, pericarp 5-7 mm thick . . . .
           . . . . 3. *C. pachyandra* Hu
        4. Bracts and sepals 7-9, stamens 1-1.5 mm long, pericarp 2-3 mm thick . . . . .
           . . . . 4. *C. xanthochroma* Feng & Xie
1. Ovaries glabrous.
    5. Branchlets glabrous.
        6. Leaves large, ca. 20 cm long.
            7. Leaf bases cordate, amplexicaul, sessile . . . . . 5. *C. amplexifolia* Merr. & Chun
            7. Leaf bases rounded, petioles 5 mm long . . . . 6. *C. brachyandra* Chang
        6. Leaves 7 cm long, 3 cm wide, flowers sessile . . . . 7. *C. nervosa* (Gagnep.) Chang
    5. Branchlets pubescent.
        8. Leaves 5-11 cm long, pedicels 2-10 mm long.
            9. Leaves elliptic, to 10 cm long, apices short-pointed or subcuneate.
                10. Flowers without bracts, pedicels ca. 3 mm long; petals 5, 1-1.3 cm long . . . . 8. *C. nematodea* (Gagnep.) Sealy
                10. Flowers with 2-3 bracts, pedicels 5-10 mm long; petals 7-8, 8 mm long . . . . 9. *C. gilbertii* (A. Chev.) Sealy
            9. Leaves oblong, 8 cm long, apices long-caudate; pedicels 2 mm long; petals 5, 1 cm long, coriaceous; bracts 2 . . . . 10. *C. crassipetala* Chang
        8. Pedicels extremely short; leaves 4-8 cm long, sometimes to almost 16 cm long.
            11. Leaves small, elliptic, 4 cm long, 2 cm wide; petals 16 mm long . . . . . .
                . . . . 11. *C. yangkiangensis* Chang
            11. Leaves orbicular, 6-8 cm long, 2-3 cm wide; petals 8 mm long . . . .
                . . . . 12. *C. parviflora* Merr. & Chun ex Sealy

1. **Camellia muricatula** Chang, Acta Sci. Nat. Univ. Sunyatseni, monogr. ser. 1: 93. 1981.

A *C. szemaoensi* differt foliis majoribus, bracteis pluribus petalis minoribus, stylis breviorbis; a *C. pachyandra* foliis angustioribus, bracteis pluribus, ramulis pubescentibus recedit.

Arbor parva circ. 8-12 m alta, ramulis puberulis. Folia tenuiter coriacea elliptica vel oblonga 6-9 cm longa 2.5-3.5 cm lata, apice abrupte acuta basi late cuneata, supra opaca subtus ad costam puberula, nervis lateralibus 7-8-jugis, margine acriter serrata, petiolis 5-7 mm longis puberulis. Flores axillares albi subsessiles; bracteis 3-4 glabris; sepalis 5 obovatis 4-5 mm longis glabris; petalis 5 ellipticis 1-1.2 cm longis; staminibus 7-8 mm longis extimis basi 3-4 mm connatis glabris; ovariis pilosis 3-locularibus, stylis 3 liberis 1 mm longis.

Small tree 8-12 m tall, branchlets puberulent. Leaves thinly coriaceous, elliptic or oblong, 6-9 cm long, 2.5-3.5 cm wide, apices acute, bases broadly cuneate, dark green above, puberulent below on the midvein, lateral veins 7-8 pairs, margins sharply serrate; petioles 5-7 mm long, pubescent. Flowers axillary, white, subsessile; bracts 3-4, glabrous; sepals 5, 4-5 mm long, obovate, glabrous; petals 5, 1-1.2 cm long, elliptic; stamens 7-8 mm long, outer filament whorl basally connate for 3-4 mm, glabrous; ovaries pubescent, 3-locular; styles 3-parted, 1 mm long. Fruit 3-locular, pubescent, immature.

Yunnan: Yingjiang Xian, Xima Commune, elev. 1700 m, mixed forests, tree, flowers white, 5 September 1972, Yang Zenghong 6918 (holotype in KUN); same loc., Xima Commune, Naban Dam, elev. 400 m, Tao Guoda 13131.

2. *Camellia szemaoensis* Chang, Acta Sci. Nat. Univ. Sunyatseni, monogr. ser. 1: 94. 1981.

A *C. pachyandra* differt ramulis pubescentibus, foliis minoribus, stylis longioribus, pericarpio tenui.

Frutex vel arbor parva, ramulis pubescentibus. Folia coriacea ovato-oblonga 4-6 cm longa 1.5-2 cm lata, supra olivaceo-viridia subtus sparse pilosa, nervis lateralibus 5-6-jugis utrinque inconspicuis, margine serrulata, petiolis 4-5 mm longis puberulis. Flores albi sessiles; bracteis 2 glabris; sepalis 5 late ovatis 5-6 mm longis leviter puberulis margine scariosis; petalis 7 obovatis circ. 1 cm longis; staminibus 8 mm longis basi connatis, tubo filamentorum brevi glabris; ovariis pilosis, stylis 3-4 mm longis liberis. Capsula globosa 2-2.5 cm in diametro 3-4-valvata dehiscens, valvis 5-7 mm crassis glabris; semina solitaria in quoque loculo.

Shrub or small tree 2 m tall, branchlets pubescent. Leaves coriaceous, obovate-oblong, 4-6 cm long, 1.5-2 cm wide, apices acuminate, bases cuneate, olive-green above in the dry state, pilose below; lateral veins 5-6 pairs, equally obscure on both surfaces; margins serrulate; petioles 4-5 mm long, pubescent. Flowers white, sessile; bracts 2, glabrous; sepals 5, broadly ovate, 5-6 mm long, puberulent, margins scarious; petals 7, 1 cm long; stamens 8 mm long, filaments forming a short tube, glabrous; ovaries pubescent, styles 3-4 mm long. Capsules globose, 2-2.5 cm in diameter, 3-4-valvate dehiscent, valves 5-7 mm thick, 1 seed per locule. Flowering period October.

Yunnan: Puwen, elev. 760-930 m, open forests in a mountain valley, Mao Pinyi 6751; Simao, Xi Shan, December 1960, Wu Zhengyi (C. Y. Wu) s.n. (holotype in KUN).

3. *Camellia pachyandra* Hu, Bull. Fan Mem. Inst. Biol. Bot. 8: 131. 1938.

C. taheishanensis Zhang, Acta Bot. Yunnanica 2: 343. 1980, *syn. nov.*

Branchlets glabrous. Leaves elliptic, submembranous, 10 cm long, 4 cm wide, midveins pubescent below. Flowers short pedicellate, bracts 1-2, sepals 4-5; petals 6, ca. 1 cm long; stamens 6 mm long, lower half connate; ovaries pubescent; styles connate, 1.5 cm long, upper half 3-parted. Capsules oblate, 4 cm wide, 2 cm high.

Yunnan: Wang Qiwu (C. W. Wang) 73204, 73288 (holotype in A); Gengma, Yunnan Investigation Team 5592; Menghai, elev. 1800 m, Zhang Fangci (F. S. Zhang) and Yu Yuexing (Z. C. Yu) 0002 (*C. taheishanensis* holotype in Yunnan Agric. Univ., isotype in KUN).

4. *Camellia xanthochroma* Feng & Xie, Acta Bot. Yunnanica 2: 424. 1980.

A *C. parviflora* persimilis, sed floribus xanthellis, ovario albo-villoso differt.

Frutex 3-4 m altus, cortice brunneus, altro-punctatus, cinereus glaber. Folia coriacea, oblonga vel oblongo-lanceolata, 6.5-12 cm longa, 2.2-3.7 cm lata, apice obtusa, interdum leviter emarginata vel mucronulata, basi cuneata vel obtusa, margine revoluta, apice ⅓ sparse valde inconspicueque denticulata, utraque 7-11-dentata, supra viridi, nitida, subtus pallidiore, minute papillosa, costa supra impressa, subtus prominenti, nervis lateralibus 7-11-jugis, supra impressis, subtus leviter prom-

1-4. *Camellia pachyandra* Hu 5-7. *Camellia pilosperma* Liang
1. sterile branch; 2. flower; 3. petals and stamens; 4. calyx and pistil;
5. flowering branch; 6. petal and stamens; 7. calyx and styles.

inentibus; petiolo 3-7 mm longo, superne excavato, glabrescentibus. Flores xanthelli, 1.2 cm diam., axillares solitarii vel terminales pedicellis 1-1.5 mm longis vel subsessilibus, glabris, bracteola sepalaque 7-9, inaequali, subrotunda, imbricata glabra, xanthalla, 3-3.5 mm longa et lata, subcoriacea, tenuiter marginata, petalis 6, late ovatis, 7 mm longis, apice rotundatis, coriaceis; staminibus 12-16, validis, carnosis, 1-1.5 mm longis, ima basi cum petala adnatis, glabris, xanthellis; ovario subgloboso, 2 mm longo, 4-3-2 loculare, albo-villoso, stylis 4-3-2, liberis, 2 mm longis glabris. Frutus parvus compresse delto-globosus, 1 cm diam., 8 mm altus, basi bracteolis sepalisque persistentibus, 3-4 valvis, valvis lignosis, 2-3 mm crassis, columna centralis 3-angulari clavata, 5-6 mm longa, apice 2-3 mm lata; semina ignota.

Shrub 3-4 m tall; bark greyish-brown, atropunctate, smooth; branchlets light grey, glabrous. Leaves coriaceous, oblong to oblong-lanceolate, 6.5-12 cm long, 2.2-3.7 cm wide; apices obtuse, sometimes emarginate or slightly mucronate; bases cuneate or obtuse; margins revolute, upper ⅓ with sparse and inconspicuous teeth, each margin with 7-11 teeth; green above, smooth; light below, smooth and glabrous, with small raised papillae; midvein sunken above, protruding below; lateral veins 7-11 pairs, slightly impressed above, slightly protruding below; petioles 3-7 mm long, canaliculate above, glabrous. Flowers small, yellow, 1.2 cm in diameter, in terminal leaf axils, rarely terminal; pedicels 1-1.5 mm long or nearly sessile, glabrous; bracts and sepals 7-9, unequal in size, overlapping, subrotund, glabrous, yellow, length and width 3-3.5 mm, thinly coriaceous, margins thin; petals 6, broadly ovate, length 7 mm, apices rounded, coriaceous; stamens 12-16, thick, fleshy, 1-1.5 mm long, adnate with petal bases, glabrous, yellow; ovaries subglobose, 2 mm long, 4-3-2-locular, white villous pubescent; styles 4-3-2, completely free, 2 mm long, glabrous. Capsules small, tricoccus, 8 mm high, 1 cm wide, with persistent bracts and sepals at base, 3-4-valvate dehiscent; pericarp woody, 2-3 mm thick; columella stick like, 3 sided, 5-6 mm long, apex 2-3 mm wide. Seeds not seen. Flowering period from December to February.

Hainan: Ya Xian, Sanlong Reservoir, elev. 160 m, 12 January 1980, Xie Lishan and Cai Ming 656 (holotype in KUN).

This species is close to *C. parviflora*, except the flowers are smaller and yellow, and the ovaries are covered with a long white tomentum.

5. *Camellia amplexifolia* Merr. & Chun, Sunyatsenia 5: 129. 1940.
   Theopsis amplexifolia (Merr. & Chun) Hu, Acta Phytotax. Sin. 8: 266. 1963.
   Shrub 4 m tall. Leaves oblong, 10-22 cm long, 3-6.5 cm wide, bases cordate, sessile, amplexicaul. Flowers small; bracts 2-3, 1-1.5 mm long; sepals 5, 2-4 mm long, ciliate; petals 5, 5-7 mm long, basally connate; stamens 3-4 mm long, arranged in 2 series, somewhat connate; ovaries glabrous, styles 3 mm long. Capsules globose or oblate, 12-13 mm wide, 3-locular, 1-2 seeds per locule.

Hainan: Baoting, Hou Kuanzhou (F. C. How) 72639 (holotype in A, isotype in SYS); 72785.

6. *Camellia brachyandra* Chang, Acta Sci. Nat. Univ. Sunyatseni, monogr. ser. 1: 95. 1981.
   Species *C. parviflorae* affinis, a qua differt ramulis glabris, foliis multo majoribus integris, petalis pluribus 7-8.
   Frutex circ. 2 m altus, ramulsi glabris. Folia anguste elliptica 14-19.5 cm longa 5.5-6.5 cm lata, apice obtusa vel acuta basi late cuneata vel subrotundata, supra opaca subtus glabra, nervis lateralibus 11-13-jugis supra visibilibus subtus leviter prominentibus, margine integra, petiolis circ. 5 mm longis glabris. Flores singuli vel bini axillares parvi, pedicellis 2-3 mm longis, bracteis 2 sub sepalis dispositis, late ovatis 2 mm longis persistentibus; sepalis 6 suborbiculatis 3-4 mm longis glabris persistentibus; petalis 7-8 obovatis exterioribus 2-3 liberis, interioribus 4-5 basi connatis; staminibus brevissimis, filamentis ad basi petalorum adnatis glabris, antheris basifixis; ovariis glabris 3-

Camellia brachyandra Chang
1. flowering branch; 2. flower.

locularibus, stylis 3, 1-2 mm longis liberis.

Shrub 2 m tall, branchlets glabrous. Leaves coriaceous, long-oblong, 14-19.5 cm long, 5.5-6.5 cm wide, apices obtuse or obtusely pointed, bases obtuse or subrounded; light green above, dark green in the dry state, not shiny, glabrous; light green below, glabrous; lateral veins 11-13 pairs, evident on both surfaces, protruding below; margins entire; petioles ca. 5 mm long, glabrous. Flowers 1-2 in terminal leaf axils, small; pedicels 2-3 mm long, glabrous; bracts 2, tightly attached to the base of the sepals, broadly ovate, 2 mm long, glabrous, persistent; sepals 6, suborbicular, 3-4 mm long, glabrous, persistent; petals 7-8, obovoid, outer 2-3 free, inner 4-5 basally slightly connate; stamens extremely short, in 2 series, lower half of the filaments adnate with the petal bases, glabrous, anthers basifixed; ovaries glabrous, 3-locular; styles 3-parted, 1-2 mm long.

Hainan: Jiaziling, elev. 150 m, mountain valley next to a small stream, 4 July 1960, SCBI Hainan Station 1039 (holotype in PE); Ya Xian, Nanlinling, Huang Zhi 34033.

7. *Camellia nervosa* (Gagnep.) Chang, Acta Sci. Nat. Univ. Sunyatseni, monogr. ser. 1: 96. 1981.

Thea nervosa Gagnep., Not. Syst. 10: 129. 1942.

Branchlets glabrous. Leaves oblong, 7 cm long, 3 cm wide. Pedicels very short; bracts and petals 6, opposite, persistent; petals 5; stamens slightly connate; ovaries glabrous, 4-locular; styles 4-cleft. Capsules obovoid, furfuraceous.

Vietnam: middle Vietnam, Col de Braian, Ht. Doinnai Prov, Poilane 24309.

As seen from Poilane 24309, sepals persistent, petals 5; free stamens, ovaries glabrous, flowers sessile etc. this species probably belongs in section *Brachyandra*.

8. *Camellia nematodea* (Gagnep.) Sealy, Rev. Gen. Camellia, 135. 1958.

Thea nematodea Gagnep., Not. Syst. 10: 129. 1942.

Tree, branchlets pubescent. Leaves elliptic, 6 cm long, 3 cm wide, chartaceous, midvein of back pubescent. Pedicels 3 mm long; bracts 1, sepals 4, equally persistent; petals 5, 1-1.3 cm long, basally connate; stamens nearly free, ovaries glabrous; styles short, 3-cleft.

Distribution: Vietnam.

9. *Camellia gilbertii* (A. Chev.) Sealy, Rev. Gen. Camellia, 136. 1958.

Thea gilbertii A. Chev., Bull. Econ. Indochine 21: 513. 1919.

Branchlets pubescent. Leaves elliptic, 10 cm long, 4.5 cm wide. Pedicels 5-10 mm long; bracts 2-3, sepals 5-6, equally persistent; petals 7-8, 8 mm long; stamens 4 mm long, basally connate; ovaries glabrous; styles 3-parted, free.

Distribution: northern Vietnam.

10. *Camellia crassipetala* Chang, Acta Sci. Nat. Univ. Sunyatseni, mongr. ser. 1: 96. 1981.

A *C. pachyandra* ramulis pilosis, petalis crassis, ovariis glabris differt; a *C. brachyandra* ramulis pilosis, foliis multo minoribus, petalis minoribus, staminibus longioribus differt.

Arbor parva, ramulis pilosis. Folia tenue coriacea oblonga 5-8 cm longa 2-3 cm lata, apice caudato-acuminata basi late cuneata, supra opace ad costam puberula, subtus primo pubescentia, mox glabrescentia, nervis lateralibus 7-8-jugis inconspicuis, margine serrulata, petiolis 3 mm longis puberulis. Flores solitarii axillare 2 cm in diametro, pedicellis 2 mm longis; bracteis 2 parvis persistentibus; sepalis 5 suborbiculatis 3-5 mm longis ciliatis; petalis 5 obovatis 1 cm longis coriaceis glabris; staminibus 4-5 mm longis, filamentis robustis basi leviter connatis glabris; ovariis glabris 3-locularibus, stylis brevissimis.

Small tree, branchlets pubescent. Leaves thinly coriaceous, oblong, 5-8 cm long, 2-3 cm wide, apices subcaudate-acuminate, bases broadly cuneate; not shiny above, midvein pubescent; at first pubescent below; lateral veins 7-8 pairs, obscure; margins serrulate; petioles 3 mm long, pubescent. Flowers in terminal leaf axils, 2 cm in diameter, pedicels 2 mm long; bracts 2, small; sepals 5, suborbicular, 3-5 mm long, ciliate; petals 5, obovoid, 1 cm long, coriaceous, glabrous; stamens 4-5 mm long, filaments thick; ovaries glabrous, 3-locular; styles extremely short.

Yunnan: Linlun, Bao Shan, elev. 2700 m, flowers white, 5 October 1938, Yu Dejun (T. T. Yü) 17900 (holotype in PE).

11. *Camellia yangkiangensis* Chang, Acta Sci. Nat. Univ. Sunyatseni, monogr. ser. 1: 97. 1981.

A *C. crassipetala* differt foliis angustioribus et longioribus, sepalis pubescentibus, petalis longioribus puberulis.

Frutex vel arbor parva circ. 2 m alta, ramulis pilosis. Folia coriacea elliptica 3-4.5 cm longa 1.5-2.3 cm lata, apice breviter acuta basi late cuneata, supra opaca ad costam puberula subtus glabra, nervis lateralibus inconspicuis, margine crenulata, petiolis 3-4 mm longis puberulis. Flores albi solitarii axillares subsessiles; bracteis 2 late ovatis 2 mm longis persistentibus; sepalis 5 late ovatis vel suborbiculatis coriacea 4-5 mm longis pubescentibus; petalis 5 late obovatis 1.4-1.8 cm longis 1.2-1.6 cm latis puberulis; staminibus 2-seriatis 5-6 mm longis, filamentis robustis basi ⅓ connatis glabris; ovariis glabris, stylis brevissimis.

Shrub or small tree 2 m tall, branchlets pubescent. Leaves coriaceous, elliptic, 3-4.5 cm long, 1.5-2.3 cm wide, apices short-acute, bases broadly cuneate; not shiny above, puberulent along the midvein; glabrous below; lateral veins obscure; margins serrulate; petioles 3-4 mm long, pubescent. Flowers white, solitary in terminal leaf axils, subsessile; bracts 2, broadly ovate, 2 mm long; sepals 5, broadly ovate or suborbicular, coriaceous, 4-5 mm long, pubescent; petals 5, broadly obovate, 1.4-1.8 cm long, 1.2-1.6 cm wide, puberulent; stamens in 2 series, 5-6 mm long; filaments thick, basal ⅓ connate, glabrous; ovaries glabrous, styles extremely short.

Yunnan: Yuanjiang, Heping Reservoir to Mufang River, sparse forests, October 1964, Li Yanhui 5727 (holotype in YNTBI).

This species is very close to *C. crassipetala* except having leaves oblong, apices long caudate, bracts pubescent, petals larger, not as coriaceous, exterior pubescent. The latter has small elliptic leaves with short-acute apices, bracts glabrous or ciliate; petals coriaceous, smaller and glabrous.

12. *Camellia parviflora* Merr. & Chun ex Sealy, Rev. Gen. Camellia, 139. 1958.

Shrub 3 m tall. Leaf size variable, 6-8(-15) cm long, 2-3(-5) cm wide. Pedicels extremely short; bracts 2-3, sepals 5, equally persistent; petals 6, short and small; stamens short, ovaries glabrous; styles 3-parted, free, 2-3 mm long. Fruit flattened tricoccus, 1-1.5 cm in diameter, pericarp thin.

Hainan: Liu Xiqi 2848; Dongfang Xian, Zeng Pei 13241, 13314; Huang Zhi 34028, 36488; Liang Xiangyue 61699, 66023; Hou Kuanzhao (F. C. How) 71994, 72580 (holotype in A, isotype in SYS).

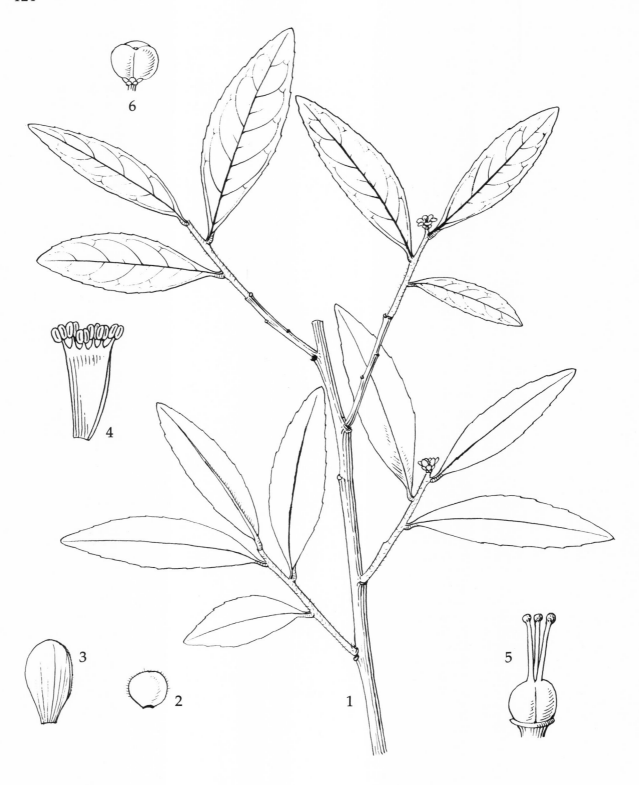

40 *Camellia parviflora* Merr. & Chun ex Sealy
1. flowering branch; 2. sepal; 3. petal; 4 stamens; 5. pistil; 6. capsule.

SECTION XIII.
Longipedicellata Chang, Acta Sci. Nat. Univ. Sunyatseni, monogr. ser. 1: 98. 1981.

Floribus terminalibus vel axillaribus solitariis mediis vel minoribus albis pedicellatis; bracteis 6-10 persistentibus; sepalis 5-7 persistentibus; petalis 8-14 leviter connatis; staminibus 2-3-seriatis filamentis exterioribus tubiformibus, ovario 3-locularii glabro, stylis 3-liberis; capsulis 3-locularibus.

Flowers 1-3 axillary or terminal, medium to large, white, pedicellate; bracts persistent, 6-10; sepals 5, persistent; petals 8-14, basally connate; stamens in 2-3 series, outer filaments connate into a tube; ovaries 3-locular, glabrous; styles 3-parted, free. Capsules 3-locular, with a columella.

4 species; 3 in Vietnam, 2 in China.

Type: *C. longipedicellata* (Hu) Chang & Fang

This section and subgenus *Metacamellia* are closely related; however, the latter has 1-locular capsules, lacking a columella, stamens in 1 series, styles connate. In the phylogeny of *Camellia* subgenus *Metacamellia* could have evolved from this section.

KEY TO SECT. *LONGIPEDICELLATA*
1. Leaves large, 10-18 cm long; petals 8 or more, 3-4 cm long.
    2. Leaves sessile, amplexicaul, basally cordate or auriculate; bracts 6-7, petals 8 . . . .
        . . . . 1. *C. amplexicaulis* (Pitard) Cohen-Stuart
    2. Leaves petiolate, bases cuneate; bracts 10, petals 14, pedicels 1 cm long . . . . .
        . . . . 2. *C. petelotii* (Merr.) Sealy
1. Leaves less than 11 cm long; petals 6-7, 1.5-3 cm long.
    3. Leaves 5 cm long, petals 3 cm long, bracts 4-6 . . . .
        . . . . 3. *C. longipedicellata* (Hu) Chang & Fang
    3. Leaves 10 cm long, thinly coriaceous; petals 1.5-2 cm long; bracts 6, opposite . . . . .
        . . . . 4. *C. indochinensis* Merr.

1. **Camellia amplexicaulis** (Pitard) Cohen-Stuart, Meded. Proefst. Thee 40: 67. 1916.
    Thea amplexicaulis Pitard in Lecomte, Fl. Gén. Indo-Chine 1: 343. 1910.
    Branchlets glabrous. Leaves elliptic, 15-16 cm long, bases auriculate. Pedicels 12 mm long, persistent; bracts 6-7; sepals 5, persistent; petals 8; filament to ⅔ length of stamens, puberulent; ovaries pubescent; styles 3-parted, free.
    Vietnam: Tien Son, T. V. Trong 3130.
    Distribution: northern Vietnam, in the region of the estuary of Yunnan River.

2. **Camellia petelotii** (Merr.) Sealy, Kew Bull. 1949(2): 219. 1949.
    Thea petelotii Merr., Univ. Calif. Publ. Bot. 10: 427. 1924.
    Branchlets glabrous. Leaves oblong, 15-18 cm long, glabrous. Pedicels 1 cm long; bracts 10, persistent; sepals 5, persistent; petals 14, stamens with a conspicuous filament tube, ovaries glabrous; styles 3-parted, free.
    Distribution: northern Vietnam.

3. **Camellia longipedicellata** (Hu) Chang & Fang, Acta Phytotax. Sin. 18: 229. 1980.
    Theopsis longipedicellata Hu, Acta Phytotax. Sin. 10: 141. 1965.
    Branchlets pubescent. Leaves elliptic, 4.5-7 cm long. Pedicels 12 mm long; bracts 4-6, persistent; sepals 7; petals 9, 3 cm long; stamens connate into a filament tube for ⅔ their length, pubescent; ovaries glabrous; styles 3-parted, free.
    Guangxi: Xincheng, flowers white, 15 February 1959, Lan Xiang 99 (holotype in PE); Douan, Liang Shengye (S. Y. Liang) 7904278.

41 1-3. *Camellia glaberrima* Chang 4-6. *Camellia longipedicellata* (Hu) Chang & Fang
1. flowering branch; 2. stamen; 3. pistil; 4. flowering branch; 5. stamens; 6. pistil.

4. *Camellia indochinensis* Merr., Journ. Arn. Arb. 20: 347. 1939.
   Thea indochinensis (Merr.) Gagnep., Suppl. Fl. Gén. Indo-Chine, 1: 307. 1943.
   Branchlets glabrous. Leaves thinly coriaceous, elliptic, 10 cm long, 4.5 cm wide. Pedicels 5-6 cm long; bracts 6, opposite; sepals 5; petals 8-9, 1.6 cm long; filament tube 8 mm long, ovaries glabrous, styles 3-parted. Capsules oblate, 4 cm wide, 2 cm high, 3-locular, 2 seeds per locule.

   Yunnan: Jinping, Mao Pinyi (P. I. Mao) 446.

   Guizhou: Luodian, Bot. Inst. Southern Quizhou Team 170, 287; Guizhou Team 496, 8867; Ceting, Cao Ziyu 1260.

   Guangxi: Tiane, Dang Baochuo 59.

   Distribution: Guizhou, Guangxi and Yunnan provinces of China as well as northern Vietnam.

Camellia indochinensis Merr.
1. flowering branch; 2. petals and stamens; 3. stamen; 4. pistil; 5. ovary cross section.

SECTION XIV.
Chrysantha Chang, Acta Sci. Nat. Univ. Sunyatseni 1979(3): 69. 1979.

Flores axillares, pedicellati; bracteis 5-7 persistentibus, sepalis 5-7 persistentibus, petalis 8-12 aureis, staminibus 4-seriatis, filamentis brevius connatis, ovariis 3-5 locularibus, stylis 3-5 liberis, ovulis 2-4 in quoque loculo.

Flowers axillary, medium large, yellow, pedicellate; bracts 5-7, persistent; sepals 5-7, persistent; petals 8-12, stamens in 4 series, filaments free or connate into a short tube; ovaries 3-5-locular, glabrous or pubescent; styles 3-5-parted, free; 2-4 seeds per locule.

10 species; 6 in Guangxi, 2 in China and Vietnam, 2 endemic to Vietnam.

Type: *C. chrysantha* (Hu) Tuyama

KEY TO SECT. *CHRYSANTHA*

1. Ovaries 5-locular, styles 5-parted . . . . Ser. I. *Flavae* Chang
   2. Leaf backs and ovaries pubescent, leaf bases slightly cordate. . . . .
      . . . . 1. *C. flava* (Pitard) Sealy
   2. Leaf backs and ovaries glabrous, leaf bases cuneate . . . . 2. *C. aurea* Chang
1. Ovaries 3-locular, styles 3-parted . . . . Ser. II. *Chrysanthae* Chang
   3. Ovaries glabrous.
      4. Flowers relatively large, 2-6 cm in diameter; fruit 1.7-4.5 cm in diameter, 1-3 locular; leaves oblong to elliptic, 6-20 cm long.
         5. Leaves coriaceous, pericarp 1-5 mm thick.
            6. Leaves oblong, 6-17 cm long, 3-5.5 cm wide.
               7. Leaves 11-16 cm long or longer, pedicels 7-10 mm long, petals to 3 cm long, fruit large, pericarp 3-5 mm thick . . . .
                  . . . . 3. *C. chrysantha* (Hu) Tuyama
               7. Leaves 6-10.5 cm long, pedicels 1-2 mm long, petals 1 cm long, fruit small, pericarp 1-1.5 mm thick . . . . 4. *C. flavida* Chang
            6. Leaves oblong, 10-20 cm long, 5-9 cm wide.
               8. Branchlets and lower surface of leaves equally pubescent, leaf veins extremely impressed . . . . 5. *C. impressinervis* Chang & Liang
               8. Branchlets and lower surfaces of leaves equally glabrous, leaf veins slightly impressed . . . . 6. *C. euphlebia* Merr. ex Sealy
         5. Leaves thinly coriaceous; pericarp thin, 1-2 mm thick.
            9. Leaves 10-15 cm long . . . . 7. *C. chrysanthoides* Chang
            9. Leaves 4-7 cm long . . . . 8. *C. tunghinensis* Chang
      4. Flowers small, 1.5-2 cm in diameter; fruit 1-1.3 cm in diameter, 1-locular; leaves ovate, less than 6.5 cm long . . . . 9. *C. pingguoensis* Fang
   3. Ovaries pubescent . . . . 10. *C. pubipetala* Wan & S. Z. Huang

SER. I.
Flavae Chang, Acta Sci. Nat. Univ. Sunyatseni 1979(3): 70. 1979.

Ovaria 5-locularia, stylis 5 liberis, staminibus subliberis.
Ovaries 5-locular; styles 5, free; stamens nearly free.
2 species in Vietnam.
Type: *C. flava* (Pitard) Sealy

1. ***Camellia flava*** (Pitard) Sealy, Kew Bull. 1949(2): 217. 1949.
   Thea flava Pitard in Lecomte, Fl. Gén. Indo-Chine 1: 346. 1910.
   Branchlets glabrous. Leaves ovate-oblong to 15 cm long, bases rounded or slightly cordate, back pubescent. Flowers pedicellate, golden-yellow; bracts 6-7; sepals 5-6, 8-15 mm long; petals 10-13, 1.5-1.8 cm long; filaments slightly connate; ovaries 5-locular,

with yellow pubescence; styles 5-parted, free, 7-15 mm long.
Distribution: northern Vietnam.

2. *Camellia aurea* Chang, Acta Sci. Nat. Univ. Sunyatseni 1979(3): 71. 1979.

A *C. flava* foliis glabris, basi late cuneatis, sepalis minoribus, petalis longioribus, ovariis stylisque glabris, petalis et stylis plus longioribus differt.

Frutex, ramulis glabris. Folia oblonga 10-15 cm longa 3.5-5 cm lata, apice subito breviter acuta, basi late cuneata vel obtusa, supra nitida subtus glabra, nervis lateralibus utrinsecus 8-9 in sicco leviter impressis, serrata, petiolis 7-9 mm longis glabris. Flores axillares solitarii, aurei, pedicellatis, pedicellis 3-5 mm longis, bracteis 5 circ. 1 mm longis, sepalis 5 subrotundatis 4-6 mm longis subglabris persistentibus, petalis 9 ellipticis vel oblongis 1.5-2.7 cm longis, staminibus 1-1.5 cm longis subliberis, ovario 5-loculari glabro, stylis 5 liberis 1.8-2.3 cm longis glabris.

Shrub, branchlets glabrous. Leaves coriaceous, oblong, 10-15 cm long, 3.5-5 cm wide, apices short-acute, bases broadly cuneate or obtuse, shiny above, glabrous below; lateral veins 8-9 pairs, sunken in the dry state; margins serrate; petioles 7-9 mm long, glabrous. Flowers solitary in leaf axils, golden-yellow; pedicels 3-5 mm long; bracts 5, 1 mm long; sepals 5, suborbicular, 4-6 mm long, glabrous, persistent; petals 9, elliptic to oblong, 1.5-2.7 cm long; stamens numerous, 1-1.5 cm long, free, glabrous; ovaries 5-locular, glabrous; styles 5-parted, free, 1.8-2.3 cm long, glabrous.

Vietnam: Liangshan Province, Youyue Xian, Exped. Sinica-Vietnamica 1599 (holotype in SCBI).

C. flava and *C. aurea* both have 5-locular ovaries and 5-parted free styles, stamens many and free. This series represents a primitive group from which ser. *Chrysantha* is derived.

SER. II.
Chrysanthae Chang, Acta Sci. Nat. Univ. Sunyatseni 1979(3): 71. 1979.

Ovaria 3-locularia, stylis 3 liberis, staminibus leviter connatis.
Ovaries 3-locular; styles 3-parted, free; stamens connate.
8 species; 6 in China, 2 in both China and Vietnam.
Type: *C. chrysantha* (Hu) Tuyama

3. *Camellia chrysantha* (Hu) Tuyama, Journ. Jap. Bot. 50: 299. 1975.
Theopsis chrysantha Hu, Acta Phytotax. Sin. 10: 139. 1965.
Camellia chrysantha var. *microcarpa* Mo & S. Z. Huang, Acta Phytotax. Sin. 17(2): 90. 1979.

Branchlets glabrous. Leaves oblong. Flowers deep yellow, pedicels 1 cm long; bracts 5, persistent; sepals 5, persistent; petals 8-10, ovaries glabrous; styles 3-4-parted, free. Capsules tricoccus.

Guangxi: Yongning, Guangxi Medicinal Inst. 17530 (holotype in PE); Gao Ruchun 17628; Nanning, Fushu Village, Liang Shengye (S. Y. Liang) and Huang Zhuomin 6403506; Dingxing, Malu Commune, Dawang Brigade, Zhong Yecong (Y. C. Chung) 621; Nanning, Zhong Yecong (Y. C. Chung) 7815, 7816; Dongying, Naliang, Qiu Huaxing 167.

Vietnam: Mone Son Hun 1g, 13 March 1962, Forestry Inst., Qujue 6173.

This species has the characteristics common in the subgenus. The size and shape of the leaves, flowers and fruit are within the range of the subgenus which make classification very difficult.

4. *Camellia flavida* Chang, Acta Sci. Nat. Univ. Sunyatseni, monogr. ser. 1: 103. 1981.

A *C. tunghinensi* differt foliis coriaceis, floribus minoribus, flavidis pedicellis brevioribus.

Frutex circ. 3 m altus, ramulis glabris. Folia coriacea elliptica vel oblonga 8-10.5 cm longa 3-4.5 cm lata, apice acuminata vel abrupte acuta, acumene obtuso, basi late cuneata, supra opaca subtus glabra, nervis lateralibus 6-7-jugis supra leviter impressis subtus prominentibus, margine serrulata, petiolis 6-8 mm longis glabris. Flores terminales flavidi, pedicellis 1-2 mm longis; bracteis 4-5 semilunatis 1.5-2.5 mm longis glabris; sepalis 5 subrobiculatis 6-8 mm longis glabris vel apicem versue plus minusve pubescentibus; petalis 8 obovatis 1 cm longis glabris; staminibus liberis glabris; ovariis glabris, stylis 3 liberis. Capsula globosa 1.7 cm in diametro 1-locularibus irregulariter 2-valvata dehiscens, valvis 1-1.5 mm crassis; semina solitaria in quoque loculo globosa 1.3 cm diam.

Shrub 3 m tall, branchlets glabrous. Leaves coriaceous, elliptic or oblong, 8-10.5 cm long, 3-4.5 cm wide, apices acuminate or abruptly acute to pointed obtuse, bases broadly cuneate; greyish-brown above in the dry state, not shiny, glabrous; light brown below, glabrous; lateral veins 6-7 pairs, slightly sunken above, protruding below; reticulating veins conspicuous below, margins serrulate; petioles 6-8 mm long, glabrous. Flowers terminal, pedicels 1-2 cm long; bracts 4-5, semiorbicular, 1.5-2.5 mm long, glabrous; sepals 5, suborbicular, 6-8 mm long, glabrous or slightly pubescent on the upper regions of the back, free; petals 8, obovoid, light yellow, ca. 1 cm long, glabrous; stamens free, glabrous; ovaries glabrous; styles 3-parted, completely free, glabrous. Capsules globose, 1.7 cm in diameter, 1-locular, 1 seeded, 2-valvate dehiscent, valves 1-1.5 mm thick; seeds globose, 1.3 cm in diameter, brown; bracts and sepals persistent.

Guangxi: Longjin, Paizong Village, mountain valley in a dense forest, flowers light yellow, 21 August 1957, Chen Shaoqing (S. C. Chun) 13736 (holotype in SCBI); without loc., Zhang Zongxiang and Wang Shanling 4095.

5. *Camellia impressinervis* Chang & Liang in Chang, Acta Sci. Nat. Univ. Sunyatseni 1979(3): 72. 1979.

Species C. *chrysanthae* affinis, a qua differt ramulis hirtellis, foliis ellipticis maximis 22 cm longis pubescentibus, nervis lateralibus 11-14-jugis, petalis 12, valvis capsularum tenuibus 1-1.5 mm crassis.

Frutex circ. 3 m altus, ramulis hirtellis. Folia coriacea elliptica 13-22 cm longa 5.5-8.5 cm lata, apice abrupte acuta, basi late cuneata vel subrotundata, supra olivaceo-viridia nitida, subtus pubescentia atropunctata, nervis lateralibus 10-14-jugis ut costa et venula impressis, margine serrata, petiolis 1 cm longis subtus pubescentibus. Flores 1-2 axillares, pedicellis robustis 6-7 mm longis glabris; bracteis 5 lunaribus dispersis glabris persistentibus; sepalis 5 semirotundatis vel orbiculatis 4-8 mm longis glabris persistentibus; petalis 12 aureis, staminibus liberis glabris; ovariis glabris, stylis 2-3 liberis glabris. Capsula oblata vel bicocca 3 cm diam. 2-3 locularis, seminibus 1-2 in quoque loculo, 1.5 cm diam.

Shrub 3 m tall; branchlets pubescent, old branches becoming bare. Leaves coriaceous, elliptic, 13-22 cm long, 5.5-8.5 cm wide, apices acute, bases broadly cuneate or slightly rounded; olive-green above in the dry state, shiny; yellowish-brown below, pubescent, dark-punctate; lateral veins 10-14 pairs; midveins and reticulating veins sunken above, protruding below; margins serrulate; petioles 1 cm long, canaliculate above, glabrous, pubescent below. Flowers 1-2 axillary, pedicells thick, 6-7 mm long, glabrous; bracts 5, crescent-shaped, oblique, glabrous, persistent; sepals 5, semi-orbicular to orbicular, 4-8 mm long, glabrous, persistent; petals 12, golden-yellow; styles 2-3-parted, glabrous. Capsules oblate or bicoccus, 1.8 cm high, 3 cm wide, 2-3-locular, 1-2 seeds per locule, valves 1-1.5 mm thick; seeds globose, 1.5 cm wide.

Guangxi: Longchuan, 13 December 1970, Liang Shengye (S. Y. Liang) 700304 (holotype in SYS); same loc., Wulian Village, Banbi Hamlet, mountain valley in a dense forest, Chen Shaoqing (S. C. Chun) 13286; Longjin, Sixth District, Banbi, Tan Peixiang 57315.

Camellia chrysantha (Hu) Tuyama 43
1. flowering branch; 2. capsules.

6. *Camellia euphlebia* Merr. ex Sealy, Kew Bull. 1949(2): 219. 1949.

Camellia chrysantha var. *macrophylla* Mo & S. Z. Huang, Acta Phytotax. Sin. 17(2): 88. 1979.

Branchlets glabrous. Leaves elliptic 11-14 cm long, 5-6.5 cm wide; on young plants leaves to 20 cm long and 9 cm wide; coriaceous, bases obtuse. Pedicels 5 mm long; bracts 7, 2-4 mm long; sepals 5, suborbicular 5-8 mm long; petals 8-9, ca. 4 cm long; stamens 3-3.5 cm long, basal half connate; ovaries glabrous; styles 3-parted, free.

Guangxi: Dongxing, Malu Commune, Dawang Brigade, Zhong Yexong (Y. C. Chung) 622.

Northern Vietnam: Near the border with Guangxi, Zeng Huaide (W. T. Tsang) 27346 (holotype in A).

Zhong 622 has leaves 20 cm long and 8.5 cm wide. Some isotypes of Zeng 27346 have leaves variable 11-20 cm long and 4.5-10 cm wide although the leaves of the holotype are shorter (to only 14.5 cm) and narrower (to only 6.5 cm).

7. *Camellia chrysanthoides* Chang, Acta Sci. Nat. Univ. Sunyatseni 1979(3): 73. 1979; Chang, Acta Sci. Nat. Univ. Sunyatseni, monogr. ser. 1: 105. 1981, *descriptio addenda*.

A *C. chrysantha* foliis membranaceis in sicco opacis, nervis lateralibus 10-11-jugis, sepalis minoribus, pericarpio tenui circ. 1 mm crasso differt.

Frutex 2.5 m altus, ramulis glabris. Folia submembranacea 10-15 cm longa 3-5.5 cm lata, apice acuminata vel abrupte acuta, basi cuneata vel leviter obtusa, supra in sicco opaca subtus glabra atropunctata, nervis lateralibus 10-11-jugis, ut costa media impressis, margine serrulata, petiolis 1 cm longis glabris. Flores axillares 4-5.5 cm diam., brevius pedicellati; bracteis 4-6; sepalis 5, subrotundatis 3-5 mm longis puberulis; petalis 8-9 basi connatis apice subacutis; staminibus 1.3-1.5 cm longis basi leviter connatis; ovario glabro, stylis 3 liberis. Capsula axillaris compresse delto-globoso vel tricocca 4.5 cm diam. 2.5 cm alta glabra 3-locularis, 3-valvata valvis tenuibus haud 1 mm crassis, ecolumnaris, seminibus 1-2 in quoque loculo, pedicellis fructiferis 6-7 mm longis, bracteis persistentibus 3-4, sepalis 5 semirotundatis vel orbicularibus 4-7 mm longis glabris.

Shrub 2.5 cm tall, branchlets glabrous. Leaves membranous, oblong or oblanceolate, 10-15 cm long, 3-5.5 cm wide, apices acuminate or abruptly acute, bases cuneate or slightly obtuse; greyish-brown above in the dry state, dull, glabrous; light brown below, glabrous, dark-punctate; lateral veins 10-11, midvein impressed above, protruding below; margins serrulate; petioles ca. 1 cm long, glabrous. Flowers axillary, 4-5.5 cm in diameter, short pedicellate; bracts 4-6; sepals 5, suborbicular, 3-5 mm long, puberulent; petals 8-9, basally slightly connate, apices slightly pointed; stamens 1.3-1.5 cm long; ovaries glabrous; styles 3-parted, free. Capsules axillary, flattened tricoccus, with 3 indented grooves, 4.5 cm wide, 2.5 cm high, glabrous, 3-locular, 1-2 seeds per locule, 3-valvate dehiscent; pericarp thin, less than 1 mm thick; with a columella; seeds globose or hemispherical, brown; pedicels 6-7 mm long; 3-4 persistent bracts, semiorbicular to orbicular, 4-7 mm long, 5-8 mm wide, glabrous.

Guangxi: Longjin, Daqing Shan, 10 September 1958, Zhang Zhaoqian 11847 (holotype in SCBI); Nanggong, Wang Bosun 7901, 7901A, 7901B, 7906, 7907, 7909; Nonggong Survey Team 11364.

8. *Camellia tunghinensis* Chang, Acta. Sci. Nat. Univ. Sunyatseni 1979(3): 73. 1979.

A speciebus ceteris differt foliis minoribus tenuibus 5-7 cm longis 2.5-3.5 cm latis, nervis lateralibus paucioribus, floribus minoribus, sepalis 4-5 mm longis, petalis 15-20 mm longis.

Frutex circ. 2 m altus, ramulis gracilibus glabris. Folia tenuiter coriacea vel submembranacea elliptica 5-7 cm longa 2.5-3.5 cm lata, apice abrupte acuta basi late cuneata, supra viridia opaca subtus atropunctata glabra, nervis lateralibus 4-5-jugis, margine serrulata, petiolis 8-15 mm longis. Flores aurei axillares, pedicellis 9-13 mm

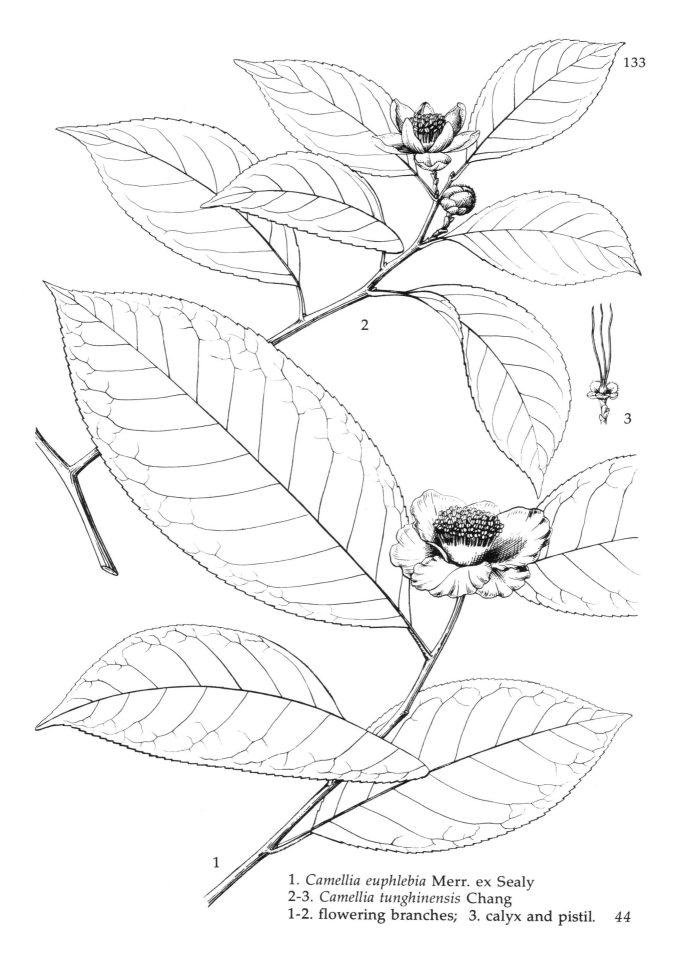

1. *Camellia euphlebia* Merr. ex Sealy
2-3. *Camellia tunghinensis* Chang
1-2. flowering branches; 3. calyx and pistil.

longis, bracteis 6-7 parvis, sepalis 5 subrotundatis 4-5 mm longis glabris, petalis 8-9, 1.5-2 cm longis basi leviter connatis, staminibus 4-5-seriatis exterioribus basi connatis, ovariis 3-locularibus glabris, stylis 3 liberis 1.5-1.8 cm longis. Capsula ignota.

Shrub 2 m tall; branchlets slender, glabrous. Leaves thinly coriaceous or nearly membranous, elliptic 5-7 cm long, 2.5-3.5 cm wide, apices acute, bases broadly cuneate; green above, not shiny; dark-punctate below, glabrous; lateral veins 4-5, upper half of margins sharply serrate, petioles 8-15 mm long. Flowers golden-yellow, axillary; pedicels 9-13 mm long; bracts 6-7, slender and small; sepals 5, suborbicular, 4-5 mm long, glabrous; petals 8-9, basally connate for 2-4 mm, 1.5-2 cm long; stamens in 4-5 series, outer whorl basally connate for 5-6 mm; ovaries glabrous, 3-locular; styles 3-parted, free, 1.5-1.8 cm long. Capsules globose, 2 cm in diameter, 1-locular, pericarp extremely thin, 2 seeded.

Guangxi: Dongxing, Neisuo, Liulan Brigade, April 1977, Yan Suzhu 77001 (holotype in SYS); Longzhou, Nonggong, Huang Qingchang s.n. (SYS nos. 149110(in fruit), 149111, 149112); Longhu, Longgong Survey Team 10249; Ningming, Long Shan, Wang Bosun 7902, 7904, 7905, 7910, 7911, 7912, 7913, 7914.

This species has the smallest leaves within the section, 5-7 cm long and 2.5-3.5 cm wide, thinly coriaceous. The flowers are also relatively small.

9. *Camellia pingguoensis* Fang, Acta Bot. Yunnanica 2: 339. 1980.

Species C. *tunghinensi* simillima, sed foliis ovatis raro lanceolatis vel anguste ellipticis, apice saepe caudatis, sepalis extus puberulis, margine ciliolatis, floribus minoribus differt.

Frutex 1-3 m altus, ramulis teretibus glabris. Folia ovata raro lanceolata vel anguste elliptica, tenuiter coriacea, 5-9.5 cm longa, 1.4-3.5 cm lata, apice abrupte acuminata saepe caudata, basi rotundato-cuneata usque cuneata raro rotundata, margine serrulata, utrinque glabra, subtus atro-punctata, costa supra leviter prominenti, nervis lateralibus ca. 6-7-jugis, supra appanatis subtus leviter elevatis, petiolo 5-10 mm longo. Flores axillares et terminales, solitarii raro geminati, aurei, tempore florendi 1.5-2.3 cm diam.; pedicelli 3-5 mm longi; bracteae ad 7, parvae; sepala 5, subrotundata, 2-3 mm longa, extus cum bracteis puberula, margine cum bracteis ciliolata, persistentia; petala 7-8, 7-13 mm longa, 6-10 mm lata, basi leviter connata; stamina 3-4-serialia, exteriora basi connata; ovarium 3-loculatum, glabrum, stylis 3, liberis, 8-12 mm longis. Capsula immatura subglobosa, ca. 1.7 cm diam., valvis tenuibus.

Shrub 1-3 m tall; branchlets glabrous, greyish-brown in the dry state. Leaves coriaceous, ovate or rarely lanceolate or thinly elliptic, 5-9.5 cm long, 1.4-3.5 cm wide, apices acute, bases broadly cuneate or nearly rounded extending below, dark green above, atropunctate below; lateral veins 6-7 pairs, slightly protruding below; margins serrulate, petioles 5-10 mm long. Flowers terminal or axillary, golden-yellow, 1.5-2.3 cm in diameter, pedicels 3-5 mm long; bracts 7, slender and small, glabrous; sepals 5, suborbicular, 2-3 mm long, ciliate; petals 7-8, 7-13 mm long, bases slightly connate; stamens in 3-4 series, 5-7 mm long, nearly free, glabrous; ovaries glabrous; styles 3-parted, free, 8-12 mm long. Capsules globose, 1.7 cm in diameter, pericarp thin, 1-3-locular, seeds very small.

Guangxi: Pingguo Xian, Pozao Commune, elev. 250 m, in secondary low forest on slope of limestone hill, flowering from October to November, Fang Ding (D. Fang) and Liao Xinpei 37692 (holotype in GXMI).

10. *Camellia pubipetala* Wan & S. Z. Huang, Acta Phytotax. Sin. 20: 316. 1982.

Species C. *impressinervi* affinis, sed bracteis, sepalis, petalis extus et filamentis pubescentibus, stylis inferne connatis superne 3-4-lobis cum ovariis pubescentibus, floribus subsessilibus differt.

Frutex vel arbuscula ad 5.5 m alta, ramulis teretibus, dense hirsutis, basi dense annulatim cicatricatis. Folia tenuiter coriacea, oblonga, oblongo-elliptica vel longe

Camellia pingguoensis Fang
1. flowering branch; 2. fruiting branch and flower bud; 3. leaf detail.   45

elliptica, 11-17 (-21) cm long, 3.5-6(-8) cm lata, apice caudato raro abrupte acuminata, apiculata, basi rotundata ad cuneata, margine serrulata, supra glabra, subtus atropunctata, dense villosa, costa media supra in sicco leviter impressa, subtus prominenti, nervis lateralibus 8-10-jugis, supra applanatis, subtus leviter prominentibus; petioli 5-10 mm longi, densi hirsuti. Flores aurei, 5-6.5 cm diam., terminales vel axillares, solitarii raro geminata, subsessiles; bracteae sepalaque 12-14, ab exterioribus ad interiores senim ampliatae, lunaria, late ovata vel subrotundata, 3-21 mm longa, extus adpresse pubescentia, margine ciliolata; petala 9-13, ovata, obovata vel oblonga, 3.2-4.5 cm longa, 1.5-2.5 cm lata, basi leviter connata, extus dense pubescentia; stamina 5-6-serialia, filamentis pubescentibus, exterioribus basi usque 1.4 cm connatis et petalis circ. 1 cm adnatis, interioribus liberis; ovarium 3-4-loculatum, subglobosum, circ. 3 mm diam., dense pubescens, stylis 2.6-3 cm longis, pubescentibus, supra medium 3-4-lobis. Capsula ignota.

Shrub or small tree, nearly 5.5 m tall; branchlets terete, densely hirsute, base with close grouped ring scars. Leaves thinly coriaceous; oblong, oblong-elliptic or long-elliptic; 11-17(-21) cm long, 3.5-6(-8) cm wide; apices caudate, rarely abruptly acuminate, apiculate; bases rounded to cuneate; margins finely serrulate; dark-green above, shiny, smooth and glabrous; yellowish-green below, dark-punctate, densely villous; midvein sunken above, protruding below; lateral veins 8-10, level on top, protruding below; petioles 5-10 mm long, densely hirsute. Flowers golden yellow, 5-6.5 cm in diameter, terminal or axillary, solitary, seldom 2, nearly sessile; bracts and sepals 12-14, gradually increasing in size from outer to inner, sublunate, broadly ovate to suborbicular, 3-21 mm long, outer surfaces puberulent, margins short ciliate; petals 9-13, ovate, obovate or orbicular, 3.2-4.5 cm long, 1.5-2.5 cm wide, bases somewhat connate, outside puberulent; stamens in many series, 5-6 whorls, filaments puberulent; outer whorl connate for 1.4 cm, adnate with petals for ca. 1 cm, inner whorl free; ovaries 3-4-locular, suborbicular, ca. 3 cm in diameter, densely puberulent; styles 2.6-3 cm long, puberulent, apices 3-4-cleft, depth of clefts ca. ⅓ the style length. Blooming period November to April. Fruit not seen.

Guangxi: Longan, in evergreen forest on slope of limestone hill, 28 November 1980, Tan Bo and Wan Yu 80042 (holotype in GXMI, isotypes in PE, SYS, SZ, IBY and GXMI); same loc., Wan Yu and Huang Zengren 80094; Huang Xiecai 7801, 7803, 7815, Wan Yu and Bi Zuzhao 81021.

This species is similar to *C. impressinervis* except that bracts, sepals, petal outer surfaces and filaments are all puberulent; style bases connate, apices 3-4-cleft; both styles and ovaries puberulent, flowers subsessile.

SECTION XV.
Calpandria (Bl.) Cohen-Stuart, Meded. Proefst. Thee 40: 69. 1916.

Calpandria Bl., Bijdr. Fl. Nederland. Indië 1: 178. 1825.

Flowers 1-2, axillary or terminal, white, relatively small; pedicels short; bracts and sepals differentiated, 4-10, persistent; petals 4-9, basally slightly connate; stamens in 2-3 series, outer filament whorl connate into a long tube; ovaries 3-locular; styles short, basally connate, apically shallowly 3-cleft.

2 species; Thailand, Malaysia, Indonesia, and the Philippines.

Type: *Camellia lanceolata* (Bl.) Seem.

KEY TO SECT. *CALPANDRIA*

1. Bracts and sepals 4-8; petals nearly free, 3-6 mm long; capsules with several seeds per locule .... 1. *C. lanceolata* (Bl.) Seem.
1. Bracts and sepals 10-11; petals basally connate with filaments for 2.5 mm, 6-8 mm long; 1 seed per locule .... 2. *C. connata* (Craib) Craib

1. **Camellia lanceolata** (Bl.) Seem., Trans. Linn. Soc. 22: 345. 1859.

Calpandria lanceolata Bl., Bijdr. Fl. Nederl. Indë. 1: 178. 1825.

Calpandria quiscosaura Korth, Kruidk., 149. 1839–1842.

Salceda montana Blanco, Fl. Filip., ed. 2, 374. 1845.

Pyrenaria camelliaeflora Vidal, Sin. Fam. y Gen. Pl. Len. Filip. Altas, 14, t. 13, fig. D. 1883, *non* Kurz

Camellia minahassae Koord., Fl. N. O. Celebes in Meded.'s Land Pl. Buitenz. 19: 350, 643. 1898.

For additional synonyms see Sealy, Rev. Gen. Camellia, 142. 1958.

Philippines: Luzon, F. Alambra s.n. (Phil. For. Dept. no. 28076); same loc., M. D. Sulit s.n. (Phil. For. Dept. no. 31080); Tayabas, M. Oro s.n. (Phil. For. Dept. no. 31056).
Celebes: Gow Lembaja, Boschbouwproefstation Buitenzorg 20888.
Distribution: Indonesia, Northern Borneo, Philippines and adjacent islands.
Branchlets pubescent. Back of leaves pubescent. Petals 4-9, filaments completely connate into a long tube; ovaries pubescent, 3-locular.

2. *Camellia connata* (Craib) Craib, Fl. Siam. Enum., 131. 1925.
Thea connata Craib, Kew Bull. 1914(1): 6. 1914.
Branchlets pubescent. Leaves oblong, 9 cm long, 4 cm wide. Pedicels 4 mm long; bracts 5-6, persistent; sepals 5, persistent; petals 5, filaments completely connate into a long tube, ovaries pubescent; styles 3-parted, extremely short. Capsules globose, 3 cm in diameter; 1 seed per locule, pericarp thin.
Distribution: Thailand.

SECTION XVI.
Thea (L.) Dyer in Hook. f., Fl. Brit. India 1: 292. 1874.

Thea L., Sp. Pl. 1: 515. 1753.
Flowers 1-3 axillary, white, medium sized or relatively small, pedicellate; bracts 2, early deciduous; sepals 5-6, persistent; petals 6-11, nearly free; stamens in 2-3 series, outer whorl nearly free, rarely connate; ovaries 3-5-locular, rarely more; styles free or lower half connate, 3-5(7) cleft. Capsules 3-5-locular, with a columella.
18 species distributed in southern and southwestern China, 2 species extending to Burma and Vietnam.
Type: *C. sinensis* (L.) O. Kuntze

KEY TO SECT. *THEA*

1. Ovaries 4-5-locular; styles 4-5-cleft, sometimes 7 cleft.
    2. Ovaries glabrous . . . . Ser. I. *Quinqueloculars* Chang
        3. Pericarp 6-8 mm thick, leaves with 9-13 pairs of lateral veins, sepals 8-10 mm long, leaves 11-17 cm long . . . . 1. *C. kwangsiensis* Chang
        3. Pericarp 2-3 mm thick, leaves with 7-12 pairs of lateral veins, sepals 5-6 mm long.
            4. Leaves coriaceous, 9-12 cm long, lateral veins 7-9 pairs, capsules globose . . . . 2. *C. quinquelocularis* Chang & Liang
            4. Leaves membranous, 12-16 cm long; lateral veins 10-12 pairs, capsules flattened tetracoccus . . . . 3. *C. tachangensis* Zhang
    2. Ovaries pubescent . . . . Ser. II. *Pentastylae* Chang
        5. Capsules globose or ovoid, ovaries 5-locular; styles 5, free, deeply cleft or 5-parted; pedicels 4-5 mm long.
            6. Capsules ovoid, columella thickened, pericarp 6-7 mm thick, flowers 6 cm in diameter, petals 9; styles deeply 5-cleft, 1.8 cm long. . . . . 4. *C. crassicolumna* Chang
            6. Capsules globose, pericarp 3-4 mm thick, flowers 4 cm in diameter, petals 12-13; styles 5-parted, free, 8-9 mm long. . . . . 5. *C. pentastyla* Chang
        5. Capsules oblate, style apices shallowly 4-5-cleft, ovaries 4-5-locular, pedicels 7-14 mm long.
            7. Leaves elliptic or narrowly elliptic, pericarp 2-3 mm thick, styles glabrous.
                8. Flowers 6 cm in diameter, pedicels 1.2-1.5 cm long, leaves broadly elliptic. . . . . 6. *C. taliensis* (W. W. Sm.) Melchior
                8. Flowers 4 cm in diameter, pedicels 8-10 mm long, leaves narrowly elliptic. . . . . 7. *C. irrawadiensis* Barua

             7. Leaves lanceolate, pericarp 4-5 mm thick, styles pubescent............
                .... 8. *C. crispula* Chang
      1. Ovaries 3-locular, styles 3-cleft or 3-parted.
          9. Ovaries glabrous............................... Ser. III. *Gymnogynae* Chang
              10. Leaves coriaceous, pedicels 7-14 mm long, sepals 3.5-6 mm long.
                  11. Bracts 5-6 mm long; leaves elliptic or narrowly elliptic, 9-13.5 cm long; pericarp 1.5-7 mm thick.
                      12. Leaves elliptic, 4-5.5 cm wide; pedicels 1 cm long; capsules large, 3-locular, valves 6-7 mm thick .... 9. *C. gymogyna* Chang
                      12. Leaves narrowly oblong or lanceolate, 2.5-3.5 cm wide, pericarp 7-8 mm long; capsules 1.5 wide, 1-locular, pericarp 1.5 mm thick ....
                          .... 10. *C. costata* Hu & Liang
                  11. Sepals 3.5 mm long; leaves oblanceolate, 8-10 cm long; pericarp 1 mm thick............................... 11. *C. yunkiangensis* Chang
              10. Leaves membranous, pedicels 4-6 mm thick, sepals 6-7 mm long........
                  .... 12. *C. leptophylla* Liang
          9. Ovaries pubescent .... Ser. IV. *Sinenses* Chang
              13. Leaf veins strongly impressed above when dry, leaves narrowly oblong, sepals 2-3 mm long................................ 13. *C. pubicosta* Merr.
              13. Leaf veins not impressed, leaves oblong or lanceolate to elliptic, sepals 3-9 mm long.
                  14. Pericarp 4-5 mm thick, leaves lanceolate, sepals 6-9 mm long........
                      .... 14. *C. angustifolia* Chang
                  14. Pericarp 1-2 mm thick, leaves elliptic or oblanceolate, sepals 3-6 mm long.
                      15. Flowers 3-4 cm in diameter, sepals 4-6 mm long.
                          16. Leaves oblong or oblanceolate, less than 10 cm long, glabrous or pubescent.
                              17. Leaves oblong, apices abruptly acute, branchlets and sepals somewhat pubescent ....
                                  .... 15a. *C. sinensis* (L.) O. Kuntze var. *sinensis*.
                              17. Leaves oblanceolate, apices caudately acuminate, branchlets and sepals glabrous...............................
                                  .... 15d. *C. sinensis* var. *waldenae* (S. Y. Hu) Chang
                          16. Leaves elliptic, 10-16 cm long, glabrous, apices acute ....
                              .... 15b. *C. sinensis* var. *assamica* (Mast.) Kitamura
                      15. Flowers small, 1.5-7.5 cm in diameter; sepals 3-5 mm long.
                          18. Leaves elliptic or oblong, lower surfaces pubescent.
                              19. Leaves membranous, less than 10 cm long...............
                                  .... 15c. *C. sinensis* var. *pubilimba* Chang
                              19. Leaves coriaceous, greater than 10 cm long.
                                  20. Sepals 5, 3-3.5 mm long; leaves 5-12.5 cm wide......
                                      .... 16. *C. fangchensis* Liang & Zhong
                                  20. Sepals 7, 4-5 mm long; leaves 4-6.5 cm wide ....
                                      .... 17. *C. ptilophylla* Chang
                          18. Leaves obovate, 11-19 cm long, lower surfaces glabrous; sepals 3 mm long .... 18. *C. parvisepala* Chang

SER. I.
Quinquelocularis Chang, Acta Sci. Nat. Univ. Sunyatseni, 1981(1): 89. 1981.
    Ovariis 4-5-locularibus glabris, stylis 4-5(7)-fidis, capsulis 4-5-locularibus.
    Ovaries 4-5-locular, glabrous; styles 4-5(-7)-cleft, capsules 4-5-locular.
    3 species distributed in southwestern China.
    Type: *Camellia quinquelocularis* Chang & Liang

1. **Camellia kwangsiensis** Chang, Acta Sci. Nat. Univ. Sunyatseni 1981(1): 89. 1981.

Species *C. pubicostae* subasimilis, a qua differt ramulis et costis foliorum glabris, sepalis majoribus circ. 1 cm longis 1.4 cm latis, fructus majoribus, pericarpio 7-8 mm crasso, ovariis subglabris 5-locularibus.

Frutex vel arbor parva, ramulis glabris. Folia coriacea oblonga 10-17 cm longa 4-7 cm lata, apice acuminata vel abrupte acuta basi late cuneata, supra opaca vel nitidula subtus glabris, nervis lateralibus 10-13-jugis a costa sub angulo 50°-60° abeuntibus, utrinque conspicuis, margine serrulata, petiolis 8-12 mm longis. Flores terminales albi, pedicellis 8 mm longis robustis; bracteis 2 caducis; sepalis 5 suborbiculatis 6-8 mm longis 8-12 mm latis extus glabris intus seriaceis; petalis et staminibus non visis; ovariis glabris 5-locularibus. Capsula globosa 2.8 cm in diametro (immatura), pericarpio 7-8 mm crasso.

Shrub or small tree, branchlets glabrous. Leaves coriaceous, oblong, 10-17 cm long, 4-7 cm wide, apices acuminate or abruptly acute, bases broadly cuneate; greyish-brown above in the dry state, not shiny or slightly glossy, glabrous; light greyish-brown below, glabrous; lateral veins 10-13 pairs, protruding slightly on both surfaces, 50°-60° angle from the midvein; reticulating veins obscure; margins serrulate, teeth separation 2-2.5 mm; petioles 8-12 mm long, glabrous. Flowers terminal; pedicels 8 mm long, thick; bracts 2, early deciduous; sepals 5, thickly coriaceous, suborbicular, 6-8 mm long, 8-12 mm wide, exterior glabrous, interior shortly sericeous; petals and stamens deciduous; ovaries glabrous, 5-locular. Capsules globose, not yet ripe, 2.8 cm in diameter, pericarp 7-8 mm thick, persistent sepals 2.5 cm in diameter.

Guangxi: Lengjiapin, in sparce forests, 28 July 1956, Li Yinkun 560 (holotype in SCBI); Tianlin, Lao Shan, hill behind Maobiliang, scattered forest, small tree, 5 November 1956, Zhang Enyuan and Li Yinkun P00719.

Yunnan: Xichou, elev. 1500-1700 m, Feng Guomei (K. M. Feng) 11580.

2. **Camellia quinquelocularis** Chang & Liang in Chang, Acta Sci. Nat. Univ. Sunyatseni 1981(1): 90. 1981.

A *C. kwangsiensi* differt foliis minoribus, nervis lateralibus paucioribus, sepalis brevioribus, capsulis minoribus, valvis tenuioribus.

Frutex vel arbor parva, ramulis glabris. Folia coriacea oblonga 9-12 cm longa, 3-4.5 cm lata, apice abrupte acuta, basi cuneata, utrinque glabra, nervis lateralibus 7-9-jugis, margine serrulata, petiolis 6-10 mm longis. Flores solitarii terminales albi, bracteis 2 caducis, pedicellis 7-9 mm longis glabris; sepalis 5 semiorbicularis 5 mm longis 7-9 mm latis glabris, petalis 12-14 obovatis circ. 2 cm longis basi connatis, staminibus 12-14 mm longis extimis basi connatis; ovariis 5-locularibus glabris, ovulis 1-4 in quoque loculo, stylis 13 mm longis apice 5-fidis. Capsula globosa 2.5 cm in diametro, 4-5-valvata dehiscens, valvis 2-2.5 mm crassis, semina globosa circ. 1 cm in diametro.

Small tree 4 m tall, branchlets glabrous. Leaves coriaceous, oblong, 9-12 cm long, 3-4.5 cm wide, apices abruptly acute, bases cuneate, glabrous, lateral veins 7-9 pairs, margins serrate, petioles 6-10 mm long. Flowers solitary, terminal, white, 3-3.5 cm in diameter; pedicels 7-9 mm long, glabrous; bracts 2, in the middle of the pedicels, early deciduous; sepals 5, suborbicular, 5 mm long, 7-9 mm wide, glabrous; petals 12-14, obovoid, 2-2.5 cm long, basally connate, glabrous; stamens 1.2-1.4 cm long, basal half of outer filament whorl connate into a short tube; ovaries 5-locular, glabrous, each locule with 1-4 ovules; styles 1.3 cm long, apically 5-cleft. Capsules globose, 2.5 cm in diameter, 4-5-valvate dehiscent, valves 2-2.5 mm thick; seeds globose, 1 cm in diameter. Flowering period November.

Guangxi: Longlin, Jinzhong Shan, elev. 1700 m, tree 4 m tall, flowers white, 23 October 1957, SCBI Geobot. Dept. 4680 (holotype in SCBI); Longlin, Liang Shengye (S. Y. Liang) 40761 (in fruit).

Yunnan: Guangnan, Muyi Commune, Huaguodaqing, Wu Quanan 9818.

46 *Camellia kwangsiensis* Chang fruiting branch.

3. *Camellia tachangensis* Zhang, Acta Bot. Yunnanica 2: 341. 1980.
Camellia tetracocca Chang, Acta Sci. Nat. Univ. Sunyatseni 1981(1): 90. 1981, *syn. nov.*

Species *C. taliensem* affinis, a qua differt foliis majoribus, 14-16(18) cm longis, 4-6(8) cm latis; floribus majoribus sub anthesi ad 6-8 cm diam., ovario glabro. A *C. irrawadiensi* foliis minoribus, 8-11.5 cm longis, 3-4 cm latis, stigmate 4-lobato, ovariis puberulis recedit.

Arbor sempervirens, 10-15 m alta, trunca ca. 40 cm crassa; cortece glauco, tenui, laevi; ramis annuis rubro-brunneis; gemma viridula, glabra. Folia coriacea tenuiora, elliptica vel longe elliptica, 14-16(-18) cm longa, 4-6(-8) cm lata, apice acuta vel caudato-acuminata, basi truncata ad rotundata, margine dilute densque serrulata, supra nitida, triste viridia, nervis lateralibus 8-9-jugis, venuis reticulatis, conspicuis; petiolis ca. 6 mm longis, glabris. Flores solitaria vel 2-3-flori, axillaria, pedicellis 1-1.5 cm longis; sepalis 5-6, rotundatis vel late ovatis, glabris; corolla alba, 6-8 cm diam., petalis 11-13(-15), late ovatis, basi connatis; staminibus numerosis, 1/5 brevioribus quam pistillo, filamentis 14 mm longis, liberatis, ad basi connatis; pistillo 1.7 cm longo, ovario supero, conico, 1-3-loculari, glabro; stigma apice retusa, basi calyce persisteni accumbens, pedicellis fructiferis ca. 2 cm longis; seminibus 1-3, subglobosis. Anthesis Aug.-Oct., ab into Oct. frutibus maturis.

Tree, 10-15 m tall, trunk diameter ca. 40 cm; bark relatively thin, greyish-white, nearly smooth; branching usually rather low and sparse, internodes 4-7 cm long, first year branches reddish-brown, strongly fragrant when cut; terminal buds pale yellow, relatively large, bud bracts imbricate, persistent, glabrous. Leaves coriaceous, relatively thin, ovate or oblong, 14-16(-18) cm long, 4-6(-8) cm wide, apices acute or caudate, bases cuneate or rounded, margins shallowly and densely serrate, leaf surfaces flat, dark green, shiny, lateral veins 8-9 pairs, reticulating veins conspicuous, midveins yellowish-green; petioles 5 mm long, glabrous. Flowers white, 6-8 cm in diameter, solitary or 2-3 in laf axils; pedicels 1-1.5 cm long; sepals 5-6, glabrous, rotundate or broadly ovate; petals 11-13(-15), basally connate, broadly ovate; stamens many, 1/5 less than the length of the pistil, filaments 14 mm long, free, basally connate; pistil 1.7 cm long, ovaries superior, conical, ca. 1.5 mm long, 3-locular, glabrous; styls 5-parted, divided for 1/5 their length. Capsules oblate, 3-4(-5) cm in diameter, ca. 2 cm high, apices indented, persistent sepals tightly attached to the base; pericarp purplish-green when ripe, rough; fruit pedicel 2 cm long, 2-4 mm thick; seeds 1-3 per locule, nearly globose, greyish-brown. Flowering period August to October. Fruit ripe the following September to October.

Yunnan: Shizong Xian, elev. 1660 m, 19 August 1979, Zhang Fangszu and Gong Yongwei 005 (holotype in Herb. Yunnan Agric. Univ., isotype in KUN).

Guizhou: Puan, Pubai Forest Center, Guizhou Agric. Purchasing Dept. 02 (holotype of *C. tetracocca* in PE).

This species is similar to *C. taliensis* except that the leaves are larger, 14-16(-18) cm long, 4-6(-8) cm wide; flowers larger, 6-8 cm in diameter; ovaries glabrous. It differs from *C. irrawadiensis* in that the latter has leaves much smaller 8-11.5 cm long, 3-4 cm wide; styles 4-parted; ovaries puberulent.

SER. II.
Pentastylae Chang, Acta Sci. Nat. Univ. Sunyatseni 1981(1): 91. 1981.

> Ovariis 4-5 locularibus pilosis, stylis 5 liberis vel apice 5-fidis.
> Ovaries 4-5-locular, pilose; styles 5, free or apices 5-cleft.
> 5 species distributed in China's southwestern province.
> Type: *C. pentastyla* Chang

4. **Camellia crassicolumna** Chang, Acta Sci. Nat. Univ. Sunyatseni 1981(1): 91. 1981.

A *C. taliensi* differt pedicellis brevioribus circ. 5 mm longis, capsulis ovoideis 4 cm longis, pericarpio crassiore 6-7 mm crasso; a *C. irrawadiensi* floribus majoribus 5-6 cm diam., petalis 3 cm longis, capsulis ovoideis 4 cm longis, pericarpio crassiore recedit.

Arbor parva circ. 10 m alta, ramulis glabris. Folia coriacea oblonga vel elliptica 10-12 cm longa 4-5.5 cm lata, apice abrupte acuta basi late cuneata, supra nitidula subtus glabra, nervis lateralibus 7-9-jugis, margine serrata petiolis 6-10 mm longis. Flores solitarii terminales 5-6 cm in diametro albi; pedicellis 5 mm longis robustis pubescentibus; bracteis 2 caducis; sepalis 5 orbiculatis 6-8 mm longis coriaceis pubescentibus; petalis 9 exterioribus 3 ovoideis 1.5 cm longis puberulis, ceteris 6 ovato-ellipticis 3 cm longis 1.5-2 cm latis puberulis basi connatis; staminibus circ. 2 cm longis subliberis glabris ovariis pilosis 5-locularibus, stylis staminibus aequantibus, apice 5-fidis. Capsula ellipsoidea 4 cm longa, 4-5-valvata dehiscens, valvis 6-7 mm crassis, columella robusta 3 cm longa 4-5-angulata, semina solitaria in quoque loculo.

Small tree 10 m tall, branchlets glabrous, terminal buds pubescent. Leaves coriaceous, oblong or elliptic, 10-12 cm long, 4-5.5 cm wide, apices acute, bases broadly cuneate, slightly shiny above, greyish tinge below, glabrous, lateral veins 7-9 pairs, margins serrate, petioles 6-10 mm long. Flowers solitary, terminal, 5-6 cm in diameter, white; pedicels 5 mm long, thick, pubescent; bracts 2, early deciduous; sepals orbicular, 6-8 mm long, coriaceous, pubescent; petals 9, outer 3 ovoid, 1.5 cm long, pubescent, remaining 6 ovate-elliptic, 3 cm long, 1.5-2 cm wide, pubescent, basally connate; stamens ca. 2 cm long, nearly free, glabrous; ovaries pubescent, 5-locular; styles as long as stamens, apically deeply 5-cleft. Capsules ovoid, 4 cm long, 4-5-valvate dehiscent, 1 seed per locule, valves 6-7 mm thick; columella large, 3 cm long, 4-5-angular. Flowering period March to April.

Yunnan: Xichou, Jian Zhuopo 644 (holotype in PE); Malipo, Feng Guomei (K. M. Feng) 13748, 13964; Masu, Doulong, Laujun Shan, Wu Quanan 8132; Pingbian, Mao Pinyi (P. I. Mao) 3107; Jinping, Li Xiwen 305; Xichou, KUN Wenshan Team 313.

5. **Camellia pentastyla** Chang, Acta Sci. Nat. Univ. Sunyatseni 1981(1): 92. 1981.

A *C. taliensi* pedicellis brevioribus, sepalis longioribus, petalis brevioribus, stylis liberis brevioribus differt; a *C. irrawadiensi* foliis majoribus, petalis pluribus, stylis liberis recedit.

Arbor circ. 10 m alta, ramulis glabris. Folia coriacea elliptica 8-12 cm longa 3.5-5.3 cm lata, apice acuta basi cuneata, supra nitida subtus glabra, nervis lateralibus 7-8-jugis, margine crenulata vel interdum subintegra, petiolis 5-10 mm longis. Flores axillares albi 4 cm in diametro, pedicellis 4-6 mm longis; bracteis 2 caducis; sepalis 5 semi-orbiculatis 4-6 mm longis glabris; petalis 12-13 basi connatis glabris; staminibus 8-10 mm longis basi leviter connatis; ovariis pilosis, stylis 5 liberis 8-9 mm longis. Capsula globosa 2.5 cm diam., semina singula in quoque loculo.

Tree 10 m tall. Leaves coriaceous, elliptic, 8-12 cm long, 3.5-5.3 cm wide, apices abruptly acute, acumine obtuse, bases broadly cuneate, shiny above, glabrous below, lateral veins 7-8 pairs, margins with small obtuse teeth or sometimes nearly entire, petioles 5-10 mm long. Flowers 4 cm in diameter, pedicels 4-6 mm long; bracts 2, early deciduous; sepals nearly semiorbicular, 4-6 mm long, outer surface glabrous; petals 12-13; stamens 8-10 mm long, basally slightly connate; ovaries pilose; styles 5-parted, completely free, 8-9 mm long. Capsules globose, 2.5 cm in diameter, 1 seed per locule.

Yunnan: Fengqing, Majie, between Yanfang and Meizhu, elev. 2050 m, 12 February 1963, Xia Lifang and Tang Zhenghong 28 (holotype in KUN), 56.

6. *Camellia taliensis* (W. W. Sm.) Melchior in Engler, Nat. Pflanzenfam., ed. 2, 21: 131. 1925.

Thea taliensis W. W. Sm., Not. Roy. Bot. Gard. Edinb. 10: 73. 1917.

Small tree, branchlets glabrous. Leaves elliptic or obovate, 6-15 cm long, 4-6 cm wide. Pedicels 1.2-1.4 cm long, bracts 2-3; sepals 5, persistent; petals 10-11, 3 cm long; stamens nearly free, ovaries pubescent, styles 4-5-cleft. Capsules oblate, 3 cm wide, 5-locular, valves 3 mm thick.

Yunnan: Dali, eastern side of Erhai, Xi Shan, G. Forrest 13477; Dali, Liu Shenno (S. N. Liu) 17919, 17923, 21958; Zhenkang, Wang Qiwu (C. W. Wang) 72680; Sino-Soviet Expedition 147.

7. *Camellia irrawadiensis* Barua, Camellian 1956(Nov.): 18. 1956.

Polyspora yunnanensis Hu, Bull. Fan Mem. Inst. Biol. Bot. 8: 135. 1938.

Gordonia yunnanensis (Hu) H. L. Li, Journ. Arn. Arb. 25: 307. 1944.

Small tree, branchlets glabrous. Leaves oblong, 12 cm long, 4.5 cm wide. Pedicels 7-8 mm long; bracts 2, deciduous; sepals 5, persistent; petals 7-10, 1.5-2 cm long, nearly free; ovaries pubescent; styles 4-5-cleft for nearly half their length, glabrous. Capsules oblate, 4 cm in diameter.

Yunnan: Wenshan Xian, elev. 2000 m, Cai Xitao (H. T. Tsai) 51511; Dali, Cai Xitao (H. T. Tsai) 56805 (holotype of *Polyspora yunnanensis*); Yuanjiang, Yin Wenqing 1669; same loc., Yang Zenghong 7530; Jingdong, Li Minggang 2025.

Distribution: China and Burma; cultivated in India.

Sealy considered Cai 51511 to be *C. taliensis,* the latter's leaves are shorter, ovate or obovate.

8. *Camellia crispula* Chang, Acta Sci. Nat. Univ. Sunyatseni 1981(1): 93. 1981.

A *C. irrawadiensi* differt foliis lanceolatis vel anguste oblongis 2-3 cm latis, capsula minore, pericarpio crassioribus, stylis pilosis.

Arbor parva, ramulis glabris. Folia tenuiter coriacea in sicco crispula lanceolata vel anguste oblonga 8-10 cm longa 2-3 cm lata, apice acuminata basi cuneata, supra opaca subtus glabra, nervis lateralibus 7-9-jugis, margine serrata, petiolis 5-7 mm longis. Flores axillares albi, pedicellis 1 cm longis; bracteis 2 caducis; sepalis 5-6 mm longis pubescentibus; petalis basi connatis; staminibus subliberis; ovariis pilosis, stylis 1 cm longis albi-pilosis, apice profunde 5-fidis. Capsula oblata 2.5 cm diam., 5-valvata dehiscens, valvis 4-5 mm crassis pilosis.

Small tree, branchlets and terminal buds glabrous. Leaves thinly coriaceous, crisped in the dry state, lanceolate or narrowly oblong, 8-10 cm long, 2-3 cm wide, apices acuminate, bases narrowly cuneate, decurrent, dull above, glabrous below, lateral veins 7-9 pairs; margins sparsely serrate, teeth separation 2-3 mm; petioles 5-7 mm long. Flowers terminal; pedicels 1 cm long, glabrous; bracts 2, early deciduous; sepals semi-orbicular, 5-6 mm long, 8-9 mm wide, pubescent; petals basally connate; stamens nearly free, basally adnate with petals; ovaries pubescent; styles 1 cm long, with white pubescence, deeply 5-cleft. Capsules oblate, 4-5-locular, 2.5 cm in diameter, with 4-5 shallow grooves, pubescent, 1 seed per locule, valves 4-5 mm thick. Flowering period March to April.

Yunnan: United Yunnan Investigation Team 1344 (holotype in KUN); Wen Shan, Laojun Shan, Feng Guomei (K. M. Feng) 22027.

SER. III.
Gymnogynae Chang, Acta Sci. Nat. Univ. Sunyatseni 1981(1): 93. 1981.

Ovariis 3-locularibus glabris, stylis apice 3-fidis.
Ovaries 3-locular, glabrous, styles 3-cleft.
4 species distributed in southern and southwestern China.
Type: *C. gymnogyna* Chang

9. *Camellia gymnogyna* Chang, Acta Sci. Nat. Univ. Sunyatseni 1981(1): 94. 1981.

Species *C. sinensem* similis, a qua differt foliis majoribus 9-13 cm longis 4-5.5 cm latis glabris, pedicellis longioribus, ovariis glabris.

Frutex, ramulis glabris. Folia coriacea elliptica 9-13.5 cm longa 4-5.5 cm lata, apice abrupte acuta basi late cuneata, supra opace subtus glabra, nervis lateralibus 8-9-jugis, margine serrata, petiolis 7-10 mm longis. Flores axillares albi, pedicellis 1 cm longis; bracteis 2 caducis; sepalis 5 late ovatis 6 mm longis glabris; petalis 7 obovatis 2 cm longis basi connatis; staminibus 1-1.2 cm longis liberis; ovariis glabris, stylis 1.2 cm longis apice 3-fidis. Capsula subglobosa 3-valvata dehiscens, valvis 3-7 mm crassis, columella 1.4 cm longa.

Shrub, branchlets glabrous, buds pubescent. Leaves coriaceous, elliptic, 9-13.5 cm long, 4-5.5 cm wide, apices acute with a 1.5 cm long acumen, bases broadly cuneate, shiny above in the dry state, glabrous below, lateral veins 8-9 pairs, margins sparsely serrate, petioles 7-10 mm long. Flowers 2, axillary; pedicels 1 cm long, robust, reflexed, glabrous; bracts 2, born on the pedicel, early deciduous; sepals 5, broadly ovate, 6 mm long, glabrous; petals 7, white, obovoid, 2 cm long, 1.2-1.5 cm wide, exterior glabrous, basally connate; stamens in many series; filaments free, 1-1.2 cm long, glabrous; ovaries glabrous; styles 1.2 cm long, apically 3-cleft, clefts 2 mm long. Capsules 3-valvate dehiscent, 1-2 locular, 1 seed per locule; valves broadly elliptic, 1.7 cm long, 3-7 mm thick; columella 1.4 cm long.

Yunnan: Xichou, Xiaoqiao Gou, Wu Quanan 62-243, 7470.

Guangxi: Lingle, Lao Shan, elev. 1400 m, 9 December 1957, Zhang Zhaoqian 11123 (holotype in SCBI); Donglan, border between Longping and Nandan, Zhang Zhaoqian 11421; Longlin, Liang Choujie 32346; Donglan, Huang Zhi (C. Wang) 43590; Qin Renchang (R. C. Ching) 6999.

Guangdong: Gaozhou, Dapo, mountains behind Yucang Village, Deng Liang 1852, 1862.

Guizhou: Leishan Xian, Yongle District, Li Junlei 03, Xixhui, Zhong Buqin (P. C. Tsoong) 296, 272.

Sichuan: Yunlain, Tuanjie Village, Sichuan Economic Plants Collecting Team 328.

10. *Camellia costata* Hu & Liang in Chang, Acta Sci. Nat. Univ. Sunyatseni 1981(1): 94. 1981.

A *C. sinensi* foliis angustioribus, pedicellis robustis, sepalis majoribus 5-6 mm longis, ovariis glabris differt.

Arbor parva, ramulis glabris. Folia anguste oblonga vel lanceolata 9-12 cm longa 2.5-3.5 cm lata, apice acuminata basi cuneata, glabra, nervis lateralibus 7-9-jugis, margine superiore serrata, petiolis 5-8 mm longis. Flores axillares, pedicellis 6-7 mm longis, bracteis 2, sepalis 5.5-6 mm longis; petalis 6-7 glabris; staminibus liberis; ovariis glabris, stylis 3-fidis. Capsula globosa 1.4 cm diam., 3-valvata dehiscens, valvis 1.5 mm crassis.

Small tree, branchlets glabrous. Leaves coriaceous, narrowly oblong or lanceolate, 9-12 cm long, 2.5-3.5 cm wide, apices acuminate, bases cuneate, slightly shiny above, yellowish-green below, glabrous, lateral veins 7-9 pairs, midvein protruding on the upper surface, upper half of the margins sparsely serrate, petioles 5-8 mm long. Flowers 1-2, axillary; pedicels 6-7 mm long, glabrous; bracts 2, on the middle of the pedicel, early deciduous; sepals 5, suborbicular, 5.5-6 mm long, bases slightly connate, glabrous; petals 6-7, glabrous; stamens nearly free; ovaries glabrous; styles glabrous, apices 3-cleft. Capsules globose, 1.4 cm in diameter, 1-locular, valves 1.5 mm thick, 1 seeded.

Guangxi: Zhaobian, Nanyong Village, Shizhu Shan, elev. 850 m, along a stream in a valley, Liang Shengye (S. Y. Liang) 6505169 (holotype in SYS).

Yunnan: Pingbian, First District, Shiban Village, Adakou, Feng Guomei (K. M. Feng) 5024.

1-3. *Camellia gymnogyna* Chang 4. *Camellia yungkiangensis* Chang
1. flowering branch; 2. stamens; 3. pistil; 4. fruiting branch.

11. *Camellia yungkiangensis* Chang, Acta Sci. Nat. Univ. Sunyatseni 1981(1): 95. 1981.

A *C. gymnogyna* differt foliis oblanceolatis, sepalis minoribus, pericarpio tenuioribus circ. 1 mm crassis.

Frutex 1.5 m altus, ramulis glabris. Folia oblanceolata 8-10 cm longa 2.5-3.5 cm lata, apice abrupte acuta vel acuminata basi cuneata, glabra, nervis lateralibus 7-8-jugis, margine serrata, petiolis 5-8 mm longis. Flores axillares albi; pedicellis 1-1.4 cm longis; bracteis 2 caducis, sepalis 5 circ. 3.5 mm longis sparse puberulis. Capsula globosa vel dicocca 2 cm diam., glabra.

Shrub 1.5 m tall, branchlets and terminal buds glabrous, older branches greyish-white. Leaves coriaceous, oblanceolate or oblong, 8-10 cm long, 2.5-3.5 cm wide, apices acute or acuminate, bases cuneate, dull above, glabrous below, lateral veins 7-8 pairs, reticulate veins obscure; margins sparsely serrate, teeth separation 4-5 mm; petioles 5-8 mm long. Flowers 1-2, axillary, pedicels 1-1.4 cm long, bracts early deciduous; sepals 3.5 cm long, sparsely pubescent. Capsules axillary, globose or bicoccus, 2 cm wide, glabrous, 2-locular, 1 seed per locule, pericarp 1 mm thick; pedicels 1.4 cm long, glabrous; bracts 2, early deciduous; sepals 5, broadly ovoid, ca. 3.5 mm long, outer surfaces glabrous, apices rounded, ciliate. Flowering period September to October.

Guizhou: Rongjiang Xian, Yueting Shan, elev. 970 m, Jian Zhuopo 51745 (holotype in PE), 51747; Bot. Inst. Qiannan Team 3149; Puan, PE Anshun Team 1362.

Yunnan: Hekou, Binglang Zhai, Kunming Work Team 5369.

Guangxi: Damiao Shan, Chen Shaoqing (S. C. Chun) 15569, 15819.

12. *Camellia leptophylla* Liang in Chang, Acta Sci. Nat. Univ. Sunyatseni 1981(1): 95. 1981.

A *C. sinensi* foliis membranaceis, petalis pluribus, ovariis glabris differt.

Frutex, ramulis pubescentibus mox glabrescentibus. Folia membranacea oblonga vel anguste elliptica 8-9.5 cm longa 3-4 cm lata, apice abrupte acuta basi cuneata, glabra, nervis lateralibus 7-8-jugis, margine serrata, petiolis 1 cm longis. Flores axillares vel terminales albi; pedicellis 4-6 mm longis glabris; sepalis 5, 6-7 mm longis ciliatis; petalis 9 obovatis 9-11 mm longis basi leviter connatis; staminibus liberis; ovariis glabris, stylis 8 mm longis apice 3-fidis.

Shrub; branchlets pubescent, soon glabrous. Leaves thinly membranous, oblong or narrowly elliptic, 8-9.5 cm long, 3-4 cm wide, apices acute with an obtuse acumen, bases broadly cuneate; dull brown above in the dry state, not shiny; light green below, glabrous; lateral veins 7-8 pairs, equally conspicuous on both surfaces; margins sparsely serrate, teeth separation 4-7 mm; petioles ca. 1 cm long, somewhat pubescent or glabrous. Flowers 1-2, terminal or axillary, white; pedicels 4-6 mm long, glabrous; bracts 2, on the middle of the pedicel, early deciduous; sepals 5, suborbicular, 6-7 mm long, glabrous, margins ciliate, persistent; petals 9, obovate, 9-11 mm long, exterior glabrous, bases slightly connate; stamens nearly free, basally adnate with petals, glabrous; ovaries glabrous; styles 8 mm long, glabrous, apically 3-cleft.

Guangxi: Longzhou, top of Daqing Shan, Liang Shenye (S. Y. Liang) 56 (GXFI no. 10551) (holotype in SYS).

SER. IV.

Sinenses Chang, Acta Sci. Nat. Univ. Sunyatseni 1981(1): 96. 1981.

>Ovariis 3-locularibus pilosis, stylis 3-fidis vel 3 liberis.
>Ovaries 3-locular, pilose; styles 3-cleft or 3-parted, free.
>6 species in southern and southwestern China.
>Type: *C. sinensis* (L.) O. Kuntze

13. *Camellia pubicosta* Merr., Journ. Arn. Arb. 23: 183. 1942.

Small tree 5-6 m tall, branchlets glabrous. Leaves narrowly oblong, 9.5-12.5 cm long, 2.5-3.5 cm wide, apices caudately acuminate, bases broadly cuneate, slightly shiny above, dark brown below in the dry state, pubescent on the small veins below; lateral veins 7-8 pairs, impressed in the dry state; margins serrulate, petioles ca. 5 mm long. Flowers solitary in the leaf axils, white; pedicels 4-5 mm long, glabrous; bracts 2, persistent; sepals 5, suborbicular, 2-3 mm long, glabrous, persistent; petals 6, basally connate, obovate, ca. 1 cm long; stamens numerous, glabrous, free, 6-7 mm long, anthers basifixed; ovaries 3-locular, pubescent; styles 3-parted, free, 7-8 mm long, puberulent.

Vietnam: Yongfu Province, Sandao, small tree, in evergreen forest, 7 February 1956, Sino-Vietnamese Investigation Team 1943.

14. *Camellia angustifolia* Chang, Acta Sci. Nat. Univ. Sunyatseni 1981(1): 96. 1981.

A *C. sinensi* ramulis et foliis glaberrimis, foliis angusto-lanceolatis, fructibus globosis, pericarpio 4-5 mm crasso, sepalis longioribus 6-9 mm longis differt.

Frutex, ramulis glabris. Folia coriacea lanceolata 7-11 cm longa 1.8-2.8 cm lata, apice acuminata basi cuneata, glabra nervis lateralibus 6-8-jugis, margine serrulata, petiolis 5-8 mm longis. Flores non visi. Capsula globosa 2.5 cm in diametro, hirsuta, 3-locularis, pericarpio 4-5 mm crasso; sepalis persistentibus 5 suborbiculatis 6-9 mm longis rotundatis glabris; pedicellis 1 cm longis.

Shrub, branchlets glabrous, older branches greyish-brown. Leaves coriaceous, lanceolate 7-11 cm long, 1.8-2.8 cm wide; apices acuminate, acumen slightly cuneate; bases obtuse, greyish-brown above in the dry state, dull, glabrous; light brown below, glabrous; lateral veins 6-8 pairs, evident on both surfaces; reticulating veins inconspicuous; margins serrulate, teeth separation 1.5-2 mm; petioles 5-8 mm long, glabrous. Flowers not seen. Capsules globose, 2.5 cm in diameter (immature), hirsute, 3-locular, pericarp 4-5 mm thick; persistent sepals 5, suborbicular, 6-9 mm long, 6-11 mm wide, apices rounded, exterior glabrous; pedicels 1 cm long, the middle has two ring scars left over from the bracts.

Guangxi: Dayao Shan, 9 July 1958, Li Yinkun 400644 (holotype in SCBI).

15. ***Camellia sinensis*** (L.) O. Kuntze., Acta Horti Petrop. 10: 1887.
   Thea sinensis L., Sp. Pl. 1: 515. 1753.
   Thea bohea L., Sp. Pl., ed. 2, 743. 1762.
   Thea viridis L., Sp. Pl., ed. 2, 735. 1762.
   Thea cantonensis Lour., Fl. Cochinch., 339. 1790.
   Thea cochinchinensis Lour., Fl. Cochinch., 338. 1790.
   Thea oleosa Lour., Fl. Cochinch., 339. 1790.
   Thea chinensis Sims, Bot. Mag. 25: sub. pl. 998. 1807.
   Camellia thea Link, Enum. Hort. Berol. 2: 73. 1822.
   Camellia sinensis var. *sinensis* f. *macrophylla* (Sieb.) Kitamura, Acta Phytotax. et Geobot. Kyoto 14: 59. 1930.
   Camellia sinensis var. *sinensis* f. *parvifolia* (Miq.) Sealy, Rev. Gen. Camellia, 116. 1958.
   For additional synonyms see Sealy, Rev. Gen. Camellia, 112-114, 116. 1958.

15a. ***Camellia sinensis*** L. var. ***sinensis***
   Anhui: Huang Shan, He Xianyu (H. Y. Ho) 2347.
   Jiangsu: Baohua Shan, A. N. Steward 2774.
   Zhejiang: Yunhe, He Xianyu (H. Y. Ho) 3732; Tianmu Shan, Shen Jun 77.
   Fujian: Tang Ruiyong 333.
   Jiangxi: Longnan, Liu Xinqi (S. K. Lau) 4724, 4308; Lu Shan, Ye Peizhong 635.

48 *Camellia angustifolia* Chang
fruiting branch.

Hunan: Wugang, Yun Shan, Zhang Hongda (H. T. Chang) 4507, 4580.
Sichuan: Emei Shan, Yang Guanghui 57500, 49718, 55679.
Xicang (Tibet): Acad. Sin. Qinghai-Xicang Team 74-1824.
Yunnan: Chen Mou (M. Chen) 3346.
Guizhou: Fangjing Shan, Jiang Ying (Y. Tsiang) 7931; Jiao Qiyuan 810.
Guangxi: Xindou, Zeng Huaide (W. T. Tsang) 23044; Guilin, Zeng Huaide (W. T. Tsang) 28358.
Guangdong: Longmen, Li Xuegen 200267; Ruyuan, Mou Ruhuai 40075.
Indonesia: Sumatra, C. Hamel 432.
Japan: K. Hisanti s.n. (SYS no. 95892).

This variety is widely distributed and commonly grown so there is a great deal of variation. The branches, leaves and flowers are either glabrous or pubescent. The leaf size is variable. Usually the leaves are smaller in the cultivated forms and the leaves are larger in the wild. The cultivated plants are shrublike and the wild plants are in the form of trees.

15b. *Camellia sinensis* var. *assamica* (Mast.) Kitamura, Acta Phytotax. et Geobot. Kyoto 14: 59. 1950.
Thea assamica Mast., Journ. Agric. et Hort. Soc. India 3: 63. 1844.
Thea viridis var. *assamica* (Mast.) Choisy, Mem. Fam. Ternst. et Camell., 67. 1855.
For additional synonyms see Sealy, Rev. Gen. Camellia, 119. 1958.
Guangxi: Zhaoping, Liang Shengye (S. Y. Liang) 6505129.
Hainan: Hongmao Shan, Zeng Huaide (W. T. Tsang) and Feng Qin (H. Feng) 18062; same loc., McClure 1858 (C.C.C. 8357), 1798 (C.C.C. 8296); Bawanling, tree 14 m tall, d.b.h. 35 cm, Zeng Pei 13448, 13374, 13841, 13339. The above are wild growing.
Guangdong: Yingde Tea Station, Zhang Hongda (H. T. Chang) 6313; Ruyuan, Zeng Pei 13841 (wild growing).
Yunnan: Longling, Kunming Work Station 203.
Vietnam: Dahuangmao Shan, Zeng Huaide (W. T. Tsang) 29277.

Regarding Hutchison's erroneous concept (The Families of Flowering Plants, 3rd ed., 334. 1973) that tea is native to India, he did not refer to Sealy (Rev. Gen. Camellia. 1958) who cited the mention of tea in early Chinese texts in his introduction. It is an indisputable fact that both *C. sinensis* var. *sinensis* and *C. sinensis* var. *assamica* are endemic to China. We also disagree with Sealy who believed that, "There can be little doubt that var. *assamica* is native in the southerly area Assam-Burma-Siam-Indo China-south China, but to what extent its distribution is due to man's agency it is now impossible to determine."

From the distribution of tea it is not possible that the natural distribution of var. *assamica* can extend to Assam. It is probable that the occurrence in Assam is due to the activities of man because the large leafed tea (*Puer*) grown presently in India was introduced from China by the East India Company (see Hutchison's description). The East India Company was established in 1600, and it was through this company that the tea growing industry was introduced to India from China. The earliest mention of tea in India is in 1826 when a member of the East India Company collected it in Assam. It was named *C. ? scottiana* by Wallich and entered into his Catalogue. After this The East India Company sent people to Assam to make a study of tea cultivation. According to the report of C. B. Bruce the tea in Assam was introduced from China by the *Dan* minority (Sealy, Rev. Gen. Camellia. 1958).

Because the cultivated plants of var. *assamica* are hardly distinguishable from the wild forms of var. *sinensis* it is probable that var. *assamica* is the original wild tea.

15c. ***Camellia sinensis*** var. ***pubilimba*** Chang, Acta Sci. Nat. Univ. Sunyatseni 1981(1): 98. 1981.

A typo differt foliis membranaceis ellipticis 5-9 cm longis 3-4 cm latis densius pubescentibus, floribus pubescentibus.

Shrub, branchlets pubescent. Leaves coriaceous, elliptic, 5-9 cm long, 3-4 cm wide, apices abruptly acute with an obtuse acumen, bases broadly cuneate; dull green above in the dry state, not shiny, somewhat tuberculate; greyish-brown below, more pubescent and less tuberculate; lateral veins 8-10 pairs, equally conspicuous on both surfaces; margins serrate; pedicels 4-5 mm long, pubescent. Flowers axillary; pedicels 4-5 mm long, abundantly pubescent; bracts and petals pubescent, stamens free, ovaries pubescent.

Guangxi: Lingyun, GXFI 4209 (holotype in SYS).

This variety differs from the typical species in that the leaves are thin, membranous, elliptic, abundantly pubescent, veins relatively more pronounced, pedicels shorter.

15d. ***Camellia sinensis*** var. ***waldenae*** (S. Y. Hu) Chang, Acta Sci. Nat. Univ. Sunyatseni 1981(1): 98. 1981.

Camellia waldenae S. Y. Hu in Wald. & S. Y. Hu, Wild Fl. Hong Kong, 61. 1977.

This variety is differentiated from the typical species in that the leaves are often oblanceolate and shiny above. The flower characters are the same as the typical species.

Hong Kong: Damao Shan, Hu Xiuying (S. Y. Hu) 12573 (holotype in A, isotypes in K, NA, PE).

Guangxi: Shangxi, Shiwan Dashan, Mo Xinli 46375; Zhaoping, Liang Shengye (S. Y. Liang) 6505244; He Xian, Liang Shengye (S. Y. Liang) 6505228.

16. ***Camellia fangchensis*** Liang & Zhong, Acta Sci. Nat. Univ. Sunyatseni 1982(3): 118. 1982.

A *C. ptylophylla* foliis tenuibus majoribus 13.5-29 cm longis et 5.5-12.5 cm latis, sepalis minoribus 3-3.5 cm longis, capsulis majoribus 1.8-3.2 cm diam. differt.

Frutex vel arbusculus 3 vel 5 m altus; cortice cinereo-brunneis; ramuli hornotini teretes atro brunnei hirsuti, annotini cinereo-brunnei glabri; gemmae pilisae. Folia subcoriacea elliptica vel oblongo-elliptica 13.5-29 cm longa et 5.5-12.5 cm lata, apice subito breviter acuta vel obtusa, basi late cuneata, supra intense virida in sicco olivaceo virida glabra, subtus viridula in sicco griseo-virida, pubescentia, nevis lateralibus 11-17-jugis ut costa elevatis; margine serrata; petiolis 3-10 mm longis supra excavatis pubescentibus. Flores parvi, albi, 2.1-3.5 cm diam., solitarii vel bini axillares, pedicellis 5-10 mm longis hirtellis; bracteis 2 deciduis, sepalis 5 suborbiculatis 3-3.5 mm longis et 2-3 mm latis extus griseo-brunneis adpresse pubescentibus, persistentibus; petalis 5 albis subrotundatis vel ovatis 10-15 mm longis et 7-10 mm latis, apice rotundatis, basi leviter connatis; extus puberulis; staminibus numerosis 3-4-seriatis, extimis basi leviter connais; interioribus basi liberis, 7.5-10 mm longis, antheris flavescentibus; ovario subgloboso 2-3 mm longo, 3 loculato, dense albo-villoso; stylo 6-10 mm longo, apice 3-fido, inferne villosulo. Capsula parva subglobosa vel compresse delto-globosa vel tricocca 1.6-2 cm alta et 1.8-3.2 cm diam. glabra 3-locularis 3-valvata valvis tenuibus haud 2 mm crassis, seminibus subglobosis convexo-angularibus solitaris in quoque loculo, 1.3-1.7 cm altis et 1.3-1.9 cm diam., pedicellis frutiferis 5-10 mm longis puberulis, sepalis persistentibus 5 subrotundatis puberulis.

Shrub or small tree, 3-5 m tall; bark greyish-brown; branchlets dark brown, terete, densely hirsute; older branches greyish-brown glabrous. Bud bracts pilose. Leaves subcoriaceous, elliptic or oblong-elliptic 13.5-29 cm long, 5.5-12.5 cm wide, apices abruptly acute or obtuse, bases broadly cuneate or rounded; dark green above, yellowish-green in the dry state; light green below, greyish-green or greyish-brown in the dry state, densely pubescent, particularly along the midvein; lateral veins 11-17 pairs, raised along with the midvein; margins serrulate; petioles 3-10 mm long, canaliculate above, pubescent. Flowers small, white, 2.1-3.5 cm in diameter, solitary or 2 born in leaf axils; pedicels 5-10 mm long, nodding, pubescent; bracts 2, early deciduous; sepals 5, suborbicular, 3-3.5 mm long, 2-3 mm wide, outer surface with greyish-brown appressed pubescence, persistent; petals 5, white, suborbicular or ovate, 10-15 mm long, 7-10 mm wide, apices rounded, bases slightly connate, exterior puberulent; stamens many, in 3-4 series, outer whorl basally slightly connate, inner whorls free, 7.5-10 mm long; anthers yellow; ovaries subglobose, 2-3 mm long, 3-locular, densely white pubescent; styles white, 6-10 mm long, apices 3-parted for ⅓ their length, bases pubescent. Capsules small, suborbicular or tricoccus oblate, with 3 sections separated by grooves, 1.6-2 cm high, 1.8-3.2 cm wide, glabrous, 3-locular, usually 1 seed per locule, 3-valvate dehiscent; pericarp thin, less than 2 mm thick; seeds subglobose, slightly angular, 1.3-1.7 cm high, 1.3-1.9 cm wide, yellowish-brown or blackish-brown; capsule pedicels 5-10 mm long, pubescent; 5 persistent sepals, subrotundate, pubescent.

Flowering period November to February, and fruit mature from November to December.

Guangxi: Fengcheng Gezu Zizhi Xian, Huashi Nawan, elev. 320 m, 3 December 1980, Zhong Yecong 80121, 80120 (holotype in GXFI); same loc., 14 January 1981, Liang Shengye (S. Y. Liang) and Zhong Yecong 8109254.

The main differences from *C. ptilophylla* are that the leaves are thinner and larger 13.5-29 cm long and 5.5-12.5 cm wide, sepals smaller 3-3.5 mm long, and capsules larger 1.8-3.2 cm in diameter.

17. *Camellia ptilophylla* Chang, Acta Sci. Nat. Univ. Sunyatseni 1981(1): 98. 1981.

A *C. sinensi* ramulis foliisque pubescentibus, foliis majoribus 12-21 cm longis, bracteis 3, sepalis 7 differt.

Arbor parva, ramulis griseo-brunneo-pubescentibus. Folia coriacea oblonga 12-21 cm longa 4-6.8 cm lata, apice acuminata basi late cuneata supra scabrida ad costam puberula, subtus pubescentia, nervis lateralibus 8-10-jugis, margine serrulata, petiolis 8-10 mm longis. Flores subterminales albi, pedicellis 8-10 mm longis, bracteis 3 caducis; sepalis 7, 4-5 mm longis puberulis; petalis 5, obovatis 1-1.2 cm longis; staminibus 8-10 mm longis liberis; ovariis 3-locularibus pilosis, stylis 1 cm longis apice 3-fidis. Capsula globosa 2 cm diam., 3-valvata dehiscens, valvis 1 mm crassis, semina singula in quoque loculo, 1.7 cm diam.

Small tree 5-6 m tall, branchlets with greyish-brown pubescence. Leaves coriaceous, 12-21 cm long, 4-6.8 cm wide; apices acuminate, acumen obtuse; bases broadly cuneate; dark green above, dull in the dry state, slightly scabrous, base of midvein pubescent; greyish-brown below in the dry state, puberulent; lateral veins 8-10, visible above, slightly protruding below; margins serrulate, teeth separation 2-4 mm; petioles 8-10 mm long, brown-pubescent. Flowers solitary, terminal; pedicels 8-10 mm long, pubescent; bracts 3, early deciduous; sepals 7, suborbicular, 4-5 mm long, exterior pubescent; petals 5, ovoid, free, 1-1.2 cm long, pubescent; stamens nearly free, 8-10 mm long, glabrous; ovaries 3-locular, pubescent; styles ca. 1 cm long, apically 3-cleft, glabrous. Capsules globose, 2 cm in diameter, pubescent, 1-3-locular, 1-2 seeds per locule, 3-valvate dehiscent, valves 1 mm thick; seeds semiglobose or globose, 1.7 cm in diameter, brown; persistent sepals 5, usually 1.2 cm wide; pedicels 1 cm long.

Guangdong: Longmen, Nankun, Zeng Pei 73, 4011 (holotype in SYS), 4012, 4013; Conghua, Lutian, Shizhang, Suiwei, mountain valley in sparse forests, small tree, 5-6 m tall, 9 November 1958, Deng Liang 8365.

18. *Camellia parvisepala* Chang, Acta Sci. Nat. Univ. Sunyatseni 1981(1): 99. 1981.

A *C. sinensi* foliis majoribus obovatis 11-19 cm longis 5-8 cm latis, floribus minoribus sepalis 3 mm longis, petalis 8-12 mm longis differt.

Frutex, ramulis pubescentibus. Folia obovata 11-19 cm longis 5-8 cm lata, apica

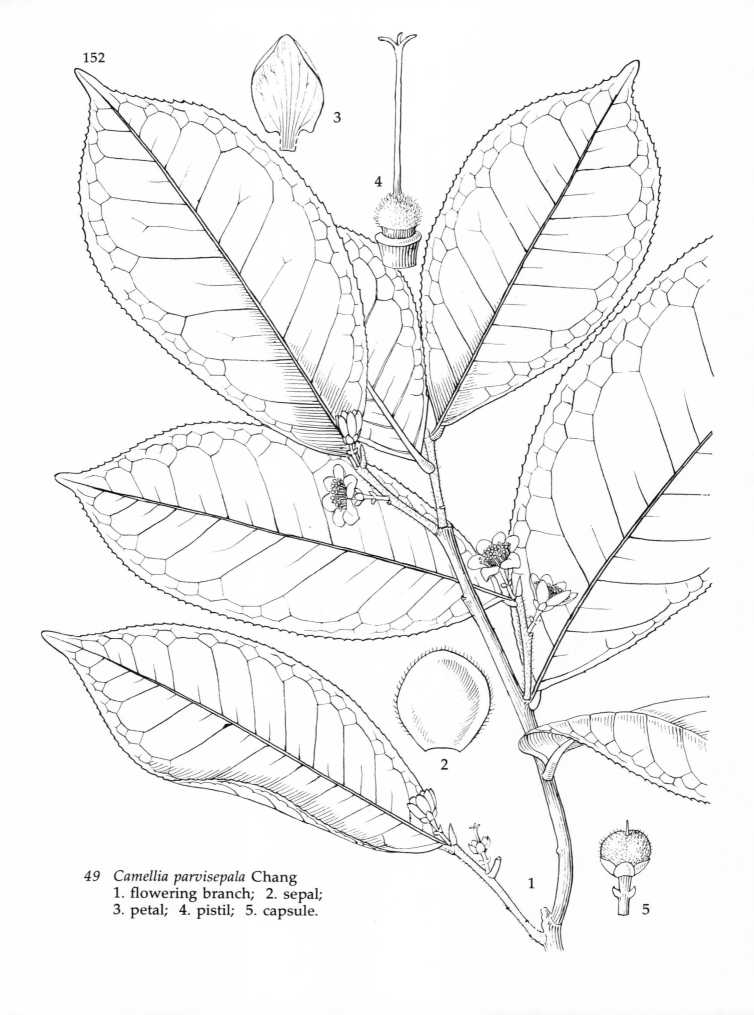

49 Camellia parvisepala Chang
1. flowering branch; 2. sepal; 3. petal; 4. pistil; 5. capsule.

abrupte acuta, nervis lateralibus 10-13-jugis, glabra, margine serrulata, petiolis 4-7 mm longis. Flores axillares pedicellis 3-5 mm longis; bracteis 2 oppositis; sepalis 5 ovoideis 3 mm longis ciliatis; petalis 6, 8-12 mm longis basi connatis; staminibus 3-4-seriatis 7-9 mm longis liberis; ovariis griseo-pilosis 3-locularibus, stylis 6 mm longis apice 3-fidis.

Shrub, branchlets pubescent. Leaves obovate, thinly coriaceous, 11-19 cm long, 5-8 cm wide, apices acute with a 1-1.5 cm long acumen, bases obtuse or slightly rounded; lateral veins 10-13 pairs, equally raised on both surfaces in the dry state; glabrous, margins serrulate, petioles 4-7 mm long. Flowers axillary, minute, white, pedicels 3-5 mm long; bracts 2, born on the middle of the pedicel, opposite; sepals 5, ovoid, 3 mm long, apices acute, ciliate; petals 6, glabrous, outer 3 petals broadly elliptic, 8-9 mm long, slightly coriaceous, inner 3 obovate, 1-1.2 cm long, basally connate; stamens 7-9 mm long, filaments free; ovaries grey-pubescent, 3-locular; styles 6 mm long, slender, glabrous, apically 3-cleft.

Guangxi: Lingle, on the road to Yuhong, 7 December 1957, Zhang Zhaoqian 11110 (holotype in SCBI); Fusui, Chen Shaoqing (S. C. Chun) 12034, 12129.

Yunnan: Simao, Manjin Shan, Feng Guomei (K. M. Feng) 14131.

SECTION XVII.
Longissima Chang, Acta Sci. Nat. Univ. Sunyatseni, monogr. ser. 1: 124. 1981.

Floribus axillaribus solitariis minoribus pedicellatis, pedicellis 4 cm longis; bracteis 2 caducis ad medium pedicellae dispositis, sepalis 5 persistentibus; petalis 7-8 subliberis; staminibus 1-2-seriatis, subliberis; ovario 3-locularii, stylis 3-fidis; capsulis a basibus versus sursum dehiscentibus.

Flowers axillary, solitary, small, pedicellate; bracts 2, early deciduous, born on the middle of the pedicel; sepals 5, persistant; petals 7-8, nearly free; stamens in 1-2 series, nearly free; ovaries 3-locular, styles 3-cleft. Capsules dehiscent from the base.

2 species; 1 in China and 1 in Vietnam.

Type: *C. longissima* Chang & Liang

KEY TO SECT. *LONGISSIMA*

1. Leaves large, elliptic, membranous, 14 pairs of lateral veins ....
   .... 1. *C. longissima* Chang & Liang
1. Leaves small, lanceolate, coriaceous, ca. 7 pairs of lateral veins ....
   .... 2. *C. gracilipes* Merr. ex Sealy

1. ***Camellia longissima*** Chang & Liang in Chang, Acta Sci. Nat. Univ. Sunyatseni, monogr. ser. 1: 124. 1981.

Foliis membranaceis, nervis lateralibus trabeculatis a costa sub angulo 80° abeuntibus, pedicellis longissimis 4 cm longis, floribus parvis, petalis et staminibus liberis, stylis brevissimis distincta.

Frutex, ramulis glabris. Folia membranacea elliptica vel obovata 14-16.5 cm longa 6-8.5 cm lata, apice abrupte acuta basi subrotundata vel late cuneata, utrinque in sicco atro-brunnea glabra, nervis lateralibus 14-19-jugis trabeculatis a costa subperpendicularibus utrinque prominentibus, retis nervorum inconspicuis, margine serrulata, petiolis 4-7 mm longis. Flores 1-3 terminales vel axillares longe pedicellati, pedicellis circ. 4 cm longis basi gracilibus apicem versus dilatatis et compressis glabris; bracteis 2 ad medium pedicelli adnatis caducis sepalis 5 late ovatis vel orbiculatis 5 mm longis glabris; petalis 8 liberis obovato-ellipticis 1 cm longis 7-8 mm latis; staminibus 7-8 mm longis liberis glabris; ovariis 3-locularibus glabris, stylis 2 mm longis apice 3-fidis.

Shrub, branchlets glabrous, old branches grey. Leaves membranous, elliptic or sometimes obovate, 14-16.5 cm long, 6-8.5 cm wide, apices abruptly acute with a ca. 1

cm long acumen, bases nearly rounded or sometimes cuneate, both surfaces dull brown in the dry state, glabrous; lateral veins 14-19 pairs, almost perpendicular to the midvein, slightly raised on both surfaces; reticulating veins not conspicuous; margins serrulate, teeth separation 2-3 mm; petioles 4-7 mm long, glabrous. Flowers 1-3, terminal or axillary, long pedicellate; pedicels ca. 4 cm long, lower region slender, upper region slightly thicker, glabrous; bracts 2, born on the middle of the pedicel, separated by 6-7 mm, early deciduous; sepals 5, broadly ovate or orbicular, 5 mm long, glabrous; petals 8, free, elliptic-obovate, ca. 1 cm long, 7-8 mm wide, glabrous; stamens 7-8 mm long, free, glabrous; ovaries 3-locular, glabrous; styles 2 mm long, apices shallowly 3-cleft.

Guangxi: Longjin, Banbi, elev. 440 m, mountain valley under dense forests, shrub, 4 October 1956, Li Zhiji 3273 (holotype in SCBI); same loc., Li Yinkun 313; same loc., GXFI 7001213.

This species can be easily identified by the membranous leaves, numerous lateral veins perpendicular to the midvein, flowers long-pedicellate, styles and petals glabrous, stamens free, ovaries 3-locular and glabrous, and styles short.

2. *Camellia gracilipes* Merr. ex Sealy, Kew Bull. 1949(2): 218. 1949.

Branchlets pubescent. Leaves narrowly lanceolate, 7-10 cm long, 1.5-2.5 cm wide. Pedicels 3-3.5 cm long, flowers small; bracts 2, opposite; sepals 5, glabrous; petals 6-7, stamens basally slightly connate, ovaries pubescent, styles 3-parted. Capsules globose, 2.5 cm in diameter.

Vietnam: Dahuangmao Shan, Zeng Huaide (W. T. Tsang) 27106, 27275 (holotype in A, isotype in SYS), 29576.

SECTION XVIII.
Glaberrima Chang, Acta Sci. Nat. Univ. Sunyatseni, monogr. ser. 1: 125. 1981.

Floribus 1-3 terminalibus axillaribusque minoribus albis pedicellatis; bracteis 2 caducis; sepalis 5 persistentibus; petalis 7-8 leviter connatis; staminibus 2-seriatis filamentis ½ inferioribus connatis tubiformibus; ovario 3-locularii, glabro, stylis 3-fidis; capsulis 3-locularibus, semine solitarii in quoque loculo.

Flowers 1-3, terminal or axillary, relatively small, white, pedicellate; bracts 2, early deciduous; sepals 5, persistent; petals 7-8, slightly connate; stamens in 2 series, basally connate into a short tube; ovaries 3-locular, glabrous; styles 3-cleft. Capsules 3-locular, 1 seed per locule.

This section is very close to sect. *Thea* differing only in that the stamens are connate into a short tube and styles are single.

2 species endemic to China.

Type: *C. glaberrima* Chang

KEY TO SECT. *GLABERRIMA*

1. Leaves large, 11-18 cm long, caudate, cauda 1-2 cm long, lateral veins conspicuous
   .... 1. *C. glaberrima* Chang
1. Leaves smaller, 9-14 cm long, apices acuminate, lateral veins obscure ....
   .... 2. *C. kwangtungensis* Chang

1. *Camellia glaberrima* Chang, Acta Sci. Nat. Univ. Sunyatseni, monogr. ser. 1: 126. 1981.

A speciebus sect. *Theae* differt staminibus alte connatis, tubis filamentorum quam filamentis liberis longioribus, foliis et floribus omnio glaberrimis.

Arbor 7-8 m alta, ramulis glabris. Folia coriacea oblonga 11-18 cm longa 4-5 cm lata, apice acuminata vel caudata, caudo 1-2 cm longo, basi cuneata vel obtusa, utrinque opaca glabra, nervis lateralibus 7-8-jugis conspicuis, margine superiore ½ serrulata, petiolis 1-1.3 cm longis. Flores solitarii terminales albi; pedicellis circ. 1 cm longis; bracteis 2 ad medium pedicelli adnatis caducis; sepalis 5 suborbicularibus 6-7 mm longis glabris persistentibus; petalis 8 obovatis circ. 2 cm longis basi leviter connatis glabris vel intus puberulis; staminibus 1.3-1.5 cm longis, extimis ⅔ connatis, intimis liberis, antheris basifixis; ovariis glabris 3-locularibus, stylis 1.8-2.2 cm longis, apice 3-fidis.

Small tree 7-8 m tall, branchlets and young leaves ca. glabrous. Leaves coriaceous, oblong, 11-18 cm long, 4-5 cm wide, apices acuminate or caudate with a 1-2 cm long cauda, bases cuneate or obtuse; upper and lower surfaces dark greyish-brown, not shiny, glabrous; lateral veins 7-8 pairs, conspicuous above, slightly protruding below; upper half of margins serrulate, teeth separation 3-5 mm; petioles 1-1.3 cm long, extremely glabrous. Flowers solitary, terminal; pedicels ca. 1 cm long, glabrous; bracts 2, born on the middle of the pedicel, early deciduous; sepals 5, suborbicular, 6-7 mm long, glabrous, persistent; petals 8, white, obovate, ca. 2 cm long, basally slightly connate, exterior glabrous, inner surface sericeous; stamens 1.3-1.5 cm long, outer filament whorl ⅔ connate, filament tube 8-9 mm long, free filaments glabrous, inner filament whorl free, anthers basifixed; ovaries glabrous, 3-locular; styles 1.8-2.2 cm long, glabrous, apices shallowly 3-cleft, clefts 1-1.3 mm long.

Yunnan: Quanping, First District, Jianshe Village, Malutang, Pujijing, elev. 1360 m, mountain valley in a sparce forest, small tree 7 m tall, flowers white, 11 March 1954, Mao Pinyi (P. I. Mao) 3330 (holotype in KUN); Jinping, Yaoshan District, Liangzizhai, small tree 8 m tall, Mao Pinyi (P. I. Mao) 3971.

2. *Camellia kwangtungensis* Chang, Acta Sci. Nat. Univ. Sunyatseni, monogr. ser. 1: 127. 1981.

A *C. glaberrima* differt foliis minoribus oblongis vel lanceolatis, sepalis et petalis minoribus.

Frutex circ. 3 m altus, ramulis glabris. Folia coriacea anguste oblonga vel lanceolata 9-14 cm longa 2.5-4 cm lata, apice caudata basi late cuneate, supra nitidula subtus glabra, nervis lateralibus 6-7-jugis utrinque inconspicuis, margine serrulata, dentis 3-5 mm remotis, petiolis 4-6 mm longis. Flores axillares albi 2 cm in diametro; pedicellis 5-6 mm longis; bracteis 2 ad medium pedicilli adnatis caducis; sepalis 5 oribuclaribus 3-4 mm longis basi leviter connatis glabris vel ciliatis; petalis 7 glabris, exterioribus 2 circ. 9-11 mm longis, interioribus 5 basi 4 mm connatis obovatis 12-15 mm longis; staminibus 1.2 cm longis 3-seriatis, extimis ⅔ connatis glabris, antheris basifixis; ovariis glabris 3-locularibus, stylis 9 mm longis apice breviter 3-fidis. Capsula compresse globosa 3 cm diam. 2 cm alta, 3-locularis, 3-valvata dehiscens, valvis 2 mm crassis; semina singula in quoque loculo.

Shrub 3 m tall, branchlets extremely glabrous. Leaves coriaceous, narrowly elliptic or lanceolate, 9-14 cm long, 2.5-4 cm wide, apices caudata, bases broadly cuneate; dark green or dull olive-green above in the dry state, slightly shiny, glabrous; concolorous below, glabrous; lateral veins 6-7 pairs, more or less obscure on both surfaces; margins sparsely serrulate, teeth separation 3-5 mm; petioles 4-6 mm long, glabrous. Flowers axillary, white, ca. 2 cm long; pedicels 5-6 mm long, glabrous; bracts 2, born on the middle of the pedicel, early deciduous; sepals 5, orbicular, 3-4 mm long, bases slightly connate, exterior glabrous, margins ciliate; petals 7, outer 2 petals 9-11 mm long, glabrous, bases slightly connate, inner 5 petals obovate, 1.2-1.5 cm long, basally connate for ca. 4 mm, glabrous; stamens 1.2 cm long, in 3 series, basal ⅔ of outer filament whorl connate, free filaments glabrous, anthers basifixed; ovaries glabrous, 3-locular; styles 9 mm long, glabrous, apices shortly 3-cleft. Capsules 3-locular, oblate, 2 cm high, 3 cm wide, 1 seed per locule, valves 2 mm thick.

Guangdong: Yang Shan, Xilu Village, Liangtang, elev. 1100 m, mountain valley in dense forests, shrub 3 m tall, flowers white, 23 November 1958, Tan Peixian 60382 (holotype in SCBI).

Camellia kwangtungensis Chang
1. sterile branch; 2. petals and stamens.

Chapter 6

SUBGENUS *Metacamellia*

SUBGEN. IV.
Metacamellia Chang, Acta Sci. Nat. Univ. Sunyatseni, monogr. ser. 1: 128. 1981.

Floribus 1-3 axillaribus terminalibusque parvis vel mediis brevius pedicellatis albis vel rarius rubris; bracteis 2-8 persistentibus; sepalis 5-6 plus minusve connatis cupuliformibus persistentibus; petalis 5-8 basi connatis; staminibus 1-2-seriatis, filamentis connatis vel subliberis; ovario 3-locularii vel rarius 4-5, stylis connatis apice 3-(-4-5) fidis; capsulis saepe 1-locularis, semine solitarii, columna nulla.

Flowers 1-3 axillary or terminal, medium sized or relatively small, white, rarely red; short pedicellate; bracts 2-8, persistent; sepals 5-6, basally somewhat connate, cup-shaped, persistent; petals 5-8, basally connate; stamens in 1-2 series, filaments connate into a tube, occasionally free; ovaries 3-locular, rarely 4-5-locular; styles connate, apices 3-cleft, occasionally 4-5-cleft. Capsules usually 1-locular, without a columella, 1 seeded.
2 sections, 57 species.
Type: *C. cuspidata* (Kochs) Hort. ex Bean

SECTION XIX.
Theopsis Cohen-Stuart, Meded. Proefst. Thee 40: 69. 1916.

Theopsis (Cohen-Stuart) Nakai, Journ. Jap. Bot. 16: 704. 1940.
Flowers 1-3, axillary or terminal, small or medium sized, white or rarely red, short pedicellate; bracts 2-8 minute, persistent; sepals 5-6, cup-shaped, persistent; petals 5-7, basally connate, stamens in 1-2 series, free or connate into a short tube, anthers dorsifixed; ovaries glabrous, styles shallowly 3-cleft. Capsules 1-locular, rarely 3-4-locular, without a columella, 1 seeded or rarely more.
This section has 43 species. Besides 1 species reaching the Ryukyu Islands the others are all found in China.
Type: *C. cuspidata* (Kochs) Hort. ex Bean

KEY TO SECT. *THEOPSIS*

1. Filaments free except for being adnate to petals . . . . Ser. I. *Cuspidatea* Chang
    2. Leaves greater than 7 cm long, width greater than 2 cm, if shorter than 7 cm the leaves are long-ovate; branchlets either pubescent or glabrous.
        3. Branchlets and flowers glabrous, leaves-long ovate.
            4. Pedicels and sepals equally pubescent, bracts 4-5.
                5. Pedicels 1.5 cm long, sepals 1 cm long . . . . . . 1. *C. macrosepala* Chang
                5. Pedicels 3-7 mm long, sepals 4-6 mm long.
                    6. Pedicels 3 mm long, sepals 4-5 mm long, petals 2-2.4 cm long. . . .
                    . . . . 2a. *C. cuspidata* (Kochs) Hort ex Bean var. *cuspidata*
                    6. Pedicels 7-8 mm long, sepals 4-6 mm long, petals 3.5-4 cm long.

            7. Leaves ovate, 7-9 cm long, apices acuminate ....
                .... 2b. *C. cuspidata* var. *grandiflora* Sealy
            7. Leaves oblong, 9-11 cm long, apices caudately acuminate, cauda 2
                cm long .... 3. *C. longicuspis* Liang
        4. Pedicels and sepals sericeous, pedicels 4-5 mm long. ....
            .... 2c. *C. cuspidata* var. *chekiangensis* Sealy
    3. Branchlets and flowers equally pubescent, bracts 2; sepals 4-6 mm long, apices
        rounded .... 4. *C. crassipes* Sealy
  2. Leaves 2-7 cm long; branchlets pubescent, rarely glabrous.
    8. Branchlets glabrous; sepals long-pointed acuminate, 9 mm long ....
        .... 5. *C. longicalyx* Chang
    8. Branchlets pubescent; sepals ovoid, apices rounded or slightly pointed.
      9. Leaves elliptic or ovate, 2-4 cm long, apices cuneate.
        10. Sepals 5 mm long, flowers to 1.8 cm long.
          11. Sepals orbicular, apices rounded ....
              .... 6a. *C. forrestii* (Diels) Cohen-Stuart var. *forrestii*
          11. Sepals ovoid, apices slightly pointed. ....
              .... 6b. *C. forrestii* var. *acutisepala* (Tsai & Feng) Chang
        10. Sepals 1-2 mm long; flowers small, 1-1.2 cm long ....
            .... 7. *C. buxifolia* Chang
      9. Leaves oblong to ovate-lanceolate, 2-7.5 cm long, apices long-pointed acute
          or caudate, rarely slightly pointed.
        12. Leaves oblong, 2-3.5 cm long, 6-9 mm wide, apices obtuse or slightly
            pointed; sepals 1-1.5 mm long. .... 8. *C. minutiflora* Chang
        12. Leaves lanceolate to ovate-lanceolate, 4-7.5 cm long, width greater than
            1 cm, apices caudate or acuminate; sepals 2-5 mm long.
          13. Leaves narrowly lanceolate, bases cuneate, sepals nearly free,
              apices acute, not ciliate; pedicels 1-2 mm long ....
              .... 9. *C. parvicuspidata* Chang
          13. Leaves ovate-lanceolate, bases rounded or obtuse; sepals free or
              connate, glabrous.
            14. Leaves 3-4.5 cm long, 1-2 cm wide; fruit pedicels 5 mm long;
                sepals basally connate, cup-shaped, 5 mm long, apices membranous, easily breaking. .... 10. *C. acutissima* Chang
            14. Leaves 5-7 cm long, 1.5-2 cm wide; fruit pedicels 1-5 mm long;
                sepals free, 2-3 mm long, coriaceous.
              15. Leaves coriaceous, obtusely serrate; fruit pedicels extremely short, sepals 2.5-3 mm long ....
                  .... 11. *C. subacutissima* Chang
              15. Leaves membranous, sharply serrate; fruit pedicels 4-5
                  mm long, sepals 2 mm long ..... 12. *C. callidonta* Chang
1. Lower half of outer filaments connate into a short tube.
  16. Filaments glabrous. .... Ser. II *Gymnandrae* Chang
    17. Sepals short and small, 1-3 mm long.
      18. Sepal outer surfaces sericeous or pubescent.
        19. Sepals long-sericeous; leaves 2-4 cm long, short-pointed ....
            .... 13. *C. handelii* Sealy
        19. Sepals puberulent, leaves 4-10 cm long.
          20. Leaves elliptic, 8-10.5 cm long; flowers in threes, pedicels
              grayish-pubescent.................... 14. *C. triantha* Chang
          20. Leaves oblong-lanceolate or ovate-oblong, 4-9 cm long; flowers
              solitary, pedicels brown-pubescent.
            21. Leaves ovate-oblong, 4-7 cm long ...... 15. *C. costei* Lév.
            21. Leaves oblong-lanceolate, to 9 cm long ....

.... 16a. *C. tsaii* Hu var. *tsaii*
18. Sepal outer surfaces glabrous or only ciliate.
   22. Leaves elliptic, ovate-oblong or lanceolate, 3-10 cm long, apices acute.
      23. Leaves 7-9 cm long, pedicels 5-6 mm long, bracts 5, sepals ciliate............. 16b. *C. tsaii* var. *synapitca* (Sealy) Chang
      23. Leaves 3-6 cm long, pedicels short or absent, bracts 4-8.
         24. Leaves ovate-lanceolate, bases subrounded.
            25. Leaves 6 cm long, flowers sessile................. .... 11. *C. subacutissima* Chang
            25. Leaves 3 cm long, pedicels 4 mm long............. .... 22b. *C. parvilimba* var. *brevipes* Chang
         24. Leaves elliptic, ovate-elliptic or lanceolate, bases cuneate; pedicels 2-4 mm long.
            26. Leaf apices acuminate, pedicels 1-4 mm long.
               27. Bracts 4.
                  28. Pedicels 2 mm long, bracts overlapping .... 17. *C. transnokoensis* Hay.
                  28. Pedicels 4 mm long, bracts separate .... 18. *C. rosthorniana* Hand.-Mazz.
               27. Bracts 8; leaves 4 cm long, midvein pubescent below.................... 19. *C. lutchuensis* Ito
            26. Leaf apices obtuse or slightly acute, pedicels 5-9 mm long....................... 20. *C. euryoides* Lindl.
   22. Leaves broadly ovate, 1-3 cm long, apices obtuse.
      29. Branchlets loosely hirsute, pedicels 2-4 mm long........... .... 21. *C. trichoclada* (Rehd.) Chien
      29. Branchlets tightly pubescent, pedicels 1 cm long............ .... 22a. *C. parvilimba* Merr. & Metc. var. *parvilimba*
17. Sepals 4-6 mm long.
   30. Sepal outer surfaces glabrous.
      31. Sepals not lanceolate, less than 8 mm long.
         32. Capsules globose or ovate, less than 2 mm long; leaves lanceolate or ovate-lanceolate.
            33. Leaves small, 4-7 cm long, 1.5-2 cm wide.
               34. Leaves ovate-lanceolate, bases nearly rounded, apices caudate; fruit pedicels 4-6 mm long................. .... 10. *C. acutissima* Chang
               34. Leaves lanceolate, bases cuneate; fruit pedicels 1 cm long, filament bases connate into a long tube .... 23. *C. elongata* (Rehd. & Wils.) Rehd.
            33. Leaves 9-11 cm long, 3-4 cm wide, caudate-acuminate; fruit pedicels 7-8 cm long .... 3. *C. longicuspis* Liang
         32. Capsules elliptic, 2.5 cm long; leaves elliptic, 6-8.5 cm long, 2-3 cm wide .... 24. *C. longicarpa* Chang ex Chang
      31. Sepals lanceolate, to 8 mm long .... 5. *C. longicalyx* Chang
   30. Sepal outer surfaces pubescent.
      36. Capsules 3-locular, pericarp 3 mm thick .... 25. *C. parvilapidea* Chang
      36. Capsules 1-locular, rarely 5-locular, pericarp 1 mm thick.
         37. Ovaries 5-locular, styles 5-cleft, capsules 1-5-locular .... 26. *C. stuartiana* Sealy
         37. Ovaries 3-locular, styles 3-cleft, capsules usually 1-locular.
            38. Sepals 4-5 mm long, bracts 3-5.

39. Leaves oblong, 3-4.5 cm long, apices obtuse; flowers 1.8 cm long, sepals yellow-villous............................
.... 27. *C. transarisanensis* (Hay.) Cohen-Stuart
39. Leaves elliptic, greater than 5 cm long, apices acuminate or caudate; flowers 2.5 cm long, sepals brown-villous or pubescent.
    40. Leaves coriaceous, 2.5-4.5 cm wide.
        41. Sepals brown-long-velutinous; pedicels 3 mm long, with pubescence .... 28. *C. fraterna* Hance
        41. Sepals brown-short-pubescent, pedicels 2-6 mm long.
            42. Pedicels 6 mm long, glabrous; petals purplish-red, leaves 6-8 cm long ....
.... 29. *C. dubia* Sealy
            42. Pedicels 2 cm long, flowers white; leaves 8-12 cm long, 3-4.5 cm wide ....
.... 30. *C. percuspidata* Chang
    40. Leaves membranous, 3-4.5 cm wide............
.... 31. *C. membranacea* Chang
38. Sepals 6-9 mm long, bracts 3-8.
    43. Bracts 6-8, sepals nearly free, petioles 6-10 mm long...................... 32. *C. rosaeflora* Hook.
    43. Bracts 3-5, sepals basally connate or cup-shaped, petioles 3-5 mm long.
        44. Calyx campanulate, 8-9 mm long, bases broad, brown-pubescent; leaves oblong; styles 3-parted, free................. 33. *C. campanisepala* Chang
        44. Calyx cup-shaped, 6 mm long, greyish-pubescent; leaves narrowly lanceolate, styles connate, apices 3-cleft................... 34. *C. lancilimba* Chang
16. Filaments pubescent........................... Ser. III. *Trichandrae* Chang
    45. Leaves ovate or long-ovate, 1.8-2.5 cm wide, bases rounded.
        46. Sepals 3 mm long, pedicels 1-2 mm long.
            47. Leaves long-ovate, sepals glabrous...........................
.... 35. *C. tsingpienensis* Hu var. *tsingpienensis*
            47. Leaves ovate, sepals pubescent ....
.... 35b. *C. tsingpienensis* var. *pubisepala* Chang
        46. Sepals 5 mm long, pedicels 4-5 mm long........................
.... 36. *C. parviovata* Chang & Wang
    45. Leaves lanceolate or oblong, 1-1.7 cm wide, bases broadly cuneate.
        48. Sepals 3-6 mm long, leaves caudate.
            49. Sepals broadly ovate, 5-6 mm long........................
.... 37. *C. viridicalyx* Chang & Liang
            49. Sepals lanceolate, 3-4 mm long.......... 38. *C. lancicalyx* Chang
        48. Sepals 1-2.5 mm long, leaf apices acute or acuminate.
            50. Flowers purplish-red, sepal exterior pubescent, leaf bases cuneate
.... 39. *C. parvicaudata* Chang
            50. Flowers white, sepals glabrous, leaf bases obtuse.
                51. Branchlets glabrous or somewhat puberulent, pedicels 1-4 mm long, sepals 2-2.5 mm long, leaves lanceolate or narrowly obovate.
                    52. Leaves lanceolate, pedicels 3-4 mm long; styles 7 mm long, pilose...................... 40. *C. subglabra* Chang
                    52. Leaves narrowly obovate; pedicels 1-2 mm long, glabrous;

                fruit flattened-tricoccus.......... 43. *C. trichandra* Chang
    51. Branchlets pubescent, pedicels 4-6 mm long, sepals 1-1.5 mm
        long, leaves oblong or ovate-oblong.
           53. Leaves oblong, bases cuneate ...... 41. *C. nokoensis* Hay.
           53. Leaves ovate-oblong, bases obtuse .... 42. *C. tsofuii* Chien

Camellia macrosepala Chang 51
branch in bud.

SER. I.
Cuspidatae Chang, Acta Sci. Nat. Univ. Sunyatseni, monogr. ser. 1: 133. 1981.

Filamenta libera glabra, ovaria glabra.
Filaments nearly free, glabrous; ovaries glabrous.
12 species, distributed south of the Changjiang (Yangtse River) in China.
Type: *Camellia cuspidata* (Kochs) Hort. ex Bean

1. *Camellia macrosepala* Chang, Acta Sci. Nat. Univ. Sunyatseni, monogr. ser. 1: 133. 1981.

Species *C. cuspidatae* affinis, a qua differt foliis majoribus, floribus majoribus sepalis ovatis 10 mm longis acutis, pedicellis 1.5 cm longis multo robustis.

Frutex vel arbor parva, ramulis glabris, gemmis pubescentibus. Folia coriacea elliptica 7-9.5 cm long 2.5-3.6 cm lata, apice caudato-acuminata, caudo 1.5-2 cm longo, basi obtusa vel subrotundata, supra nitida subtus glabra, nervis lateralibus 7-9-jugis utrinque inconspicuis, margine serrulata, petiolis 4-6 mm longis supra puberulis. Flores axillares albi, pedicellis 1.5 cm longis robustis glabris; bracteis 4-5 longe ovatis maximis 5 mm longis; sepalis ovatis 1 cm longis basi 2.5 mm connatis acuta glabris; petalis 7 obovatis basi connatis glabris; staminibus 2-3-seriatis subliberis; ovariis glabris 3-locularibus, stylis apice 3-fidis.

Shrub or small tree, branchlets glabrous, terminal buds pubescent. Leaves coriaceous, elliptic, 7-9.5 cm long, 2.5-3.6 cm wide, apices caudate-acuminate with a 1.5-2 cm long cauda, bases obtuse or subrounded; olive-green above in the dry state, shiny; same color below, glabrous; lateral veins 7-9 pairs, obscure on both surfaces; margins serrate; petioles 4-6 mm long, pubescent above. Flowers axillary; pedicels 1.5 cm long, glabrous, extremely robust; bracts 4-5, long-ovate, to 5 mm long; sepals 5, ovate, 1 cm long, glabrous, basal 2.5 mm connate, apices acute; petals 7, glabrous, obovate, not open, bases connate; stamens glabrous, outer stamens basally adnate with petals, others free; ovaries glabrous, 3-locular; styles apically 3-cleft.

Guangdong: Beijiang, Chen Huanyong (W. Y. Chun) 5935 (holotype in SCBI).

This species is very close to *C. cuspidata* only the leaves are larger, flowers larger, sepals 1 cm long, pedicels 2.5 cm long and extremely robust.

2. *Camellia cuspidata* (Kochs) Hort. ex Bean, Trees and Shrubs Brit. Isl. 1: 284. 1914; Hort., Gard. Chron. ser. 3, 61: 228. 1912.
  Thea cuspidata Kochs in Engler, Bot. Jarhb. 27: 586. 1900.
  Thea rosaeflora var. *glabra* Kochs in Engler, Bot. Jarhb. 27: 472. 1900.

2a. *Camellia cuspidata* (Kochs) Hort. ex Bean var. *cuspidata*

Branchlets glabrous. Leaves ovate-elliptic, 5-8 cm long. Flower parts glabrous, pedicels 3-4 mm long, bracts 4; sepals 4-5.5 mm long, coriaceous; stamens free, ovaries glabrous.

Shaanxi: Zhenbai, Department of Forestry, Northwestern Institute of Agriculture, Wang Mingchang 1193, 901.

Sichuan: Nanchuan, Jinfu Shan, Li Guofeng 63590; Nanchuan, Xiongjihua, Zhou Zilin 90489.

Yunnan: Wujilu, PE 557792.

Guangxi: Longsheng, Yuan Shufen and Liu Lanfang 5144; Guilin, Zeng Huaide (W. T. Tsang) 28382.

Guangdong: Pingyuan, Deng Liang 4133, 4352; Fuyuan, Wuzhi Shan, Zeng Pei 13637, 13892, 13885; Renhua, Zeng Huaide (W. T. Tsang) 26394; Fuyuan, Tianjing Shan, Huang Zhi (C. Wang) 42224.

Distribution: Shaanxi, Sichuan, Yunnan, Hubei, Anhui, Hunan, Guizhou, Guangxi, Guangdong, Jiangxi, Fujian.

2b. ***Camellia cuspidata*** var. ***grandiflora*** Sealy, Kew Bull. 1950(2): 216. 1950.
   Hunan: Wugang, Yun Shan, Zhang Hongda (H. T. Chang) 4525.
   Jiangxi: Anfu, Mingyueh Shan, Zhou Xixiang and Zhang Chunhua 01, 02.
   Sichuan: Yang Xianpu 3070.
   Zhang 4525 above is from the type locality of this variety. The leaf shape is exactly the same as the type but it is in fruit. The Jiangxi specimen, Zhou 01 and 02 have petals 2-2.5 cm long.

2c. ***Camellia cuspidata*** var. ***chekiangensis*** Sealy, Rev. Gen. Camellia, 58. 1958.
   Zhejiang: Tiantai, Qin Renchang (R. C. Ching) 1479 (holotype in K, isotype in PE); Xianju, Qin Renchang (R. C. Ching) 1651; Longquan, Zhang Shaoyao 7291; Taishu, Zhang Shaoyao 3659, 3690.
   This variety is differentiated from the type in that the bracts and sepals are more or less yellow-pubescent, filaments basally semiconnate.

3. ***Camellia longicuspis*** Liang in Chang, Acta Sci. Nat. Univ. Sunyatseni, monogr. ser. 1: 136. 1981.
   A *C. cuspidata* differt foliis longioribus, apice caudatis, pedicellis longioribus circ. 7-8 mm longis, fructibus minoribus.
   Ramulis glabris, foliis tenuiter coriaceis longicaudatis, pedicellis 7-8 mm longis, bracteis 4, sepalis 4 mm longis connatis rotundatis glabris, capsulis glabris 1-locularibus.
   Small tree 4 m tall; branchlets slender, glabrous. Leaves thinly coriaceous, oblong, 9-11 cm long, 3-4 cm wide, apices caudate-acuminate with a 2 cm long cauda, bases cuneate; green above, shiny; glabrous below; lateral veins 7 8 pairs, inconspicuous; margins serrulate, petioles 5-7 mm long. Capsules ovoid, 1.4 cm long, 1 cm wide, glabrous, valves thin, 1-locular, 1 seeded; pedicels 7-8 mm long, glabrous; bracts 4, ovate, 0.5-1 mm long, ciliate, persistent; sepals 4 mm long, basal half connate, glabrous, apices rounded.
   Guangxi: Rongshui Xian, Qiujian Shan, elev. 1000 m, mountain valley in forests, small tree 4 m tall, 12 August 1972, Liang Shengye (S. Y. Liang) 72029 (holotype in SYS).
   This species is very close to *C. percuspidata* only the pedicel of the fruit is longer, sepals glabrous; in addition a flowering specimen has not been seen.

4. ***Camellia crassipes*** Sealy, Kew Bull. 1949(2): 215. 1949.
   Branchlets pubescent. Leaves elliptic 6-10 cm long, 2-3 cm wide, caudate. Pedicels thick and short, bracts 2, sepals pilose, petals broadly obovate, stamens free, ovaries glabrous.
   Yunnan: Jinping, elev. 2000 m, Wu Sugong (S. K. Wu) 3853. Compared with the type this specimen has larger leaves with a longer cauda.

5. ***Camellia longicalyx*** Chang, Acta Sci. Nat. Univ. Sunyatseni, monogr. ser. 1: 136. 1981.
   C. acutisepala Sealy, Rev. Gen. Camellia, 61. 1958, *auct. non* Tsai & Feng.
   A *C. forrestii* var. *acutisepala* differt ovato-lanceolatis, bracteis lanceolatis 4 mm longis, sepalis longe lanceolatis circ. 8 mm longis acuminatis.
   Arbor circ. 4 m alta, ramulis glabris. Folia coriacea ovatolanceolata 4-6 cm longa 1.3-2 cm lata, apice acuminata, basi late cuneata, supra in sicco opaca ad costam primo puberula demum glabrescentia, subtus glabra, nervis lateralibus 5-6-jugis, margine serrata, petiolis 3-5 mm longis. Flores solitarii subterminaliter axillares albi, pedicellis 8 mm longis; bracteis 4 lanceolatis 4-5 mm longis; sepalis 8 mm longis basi 2 mm connatis, apice acuminatis glabris; petalis 5; staminibus 19; ovariis glabris, stylis apice 3-fidis.

Camellia longicuspis Liang
fruiting branch.

Camellia longicalyx Chang
1. flowering branch; 2. flower;
3. petals and stamens; 4. pistil.

Small tree 4 m tall; branchlets glabrous. Leaves coriaceous, ovate-lanceolate, 4-6 cm long, 1.3-2 cm wide, apices acuminate, bases broadly cuneate; not shiny above, midveins at first pubescent; glabrous below, lateral veins 5-6 pairs, margins serrate, petioles 3-5 mm long. Flowers solitary in terminal leaf axils, white, pedicels 8 mm long; bracts 4, lanceolate, 4-5 mm long; sepals 8 mm long, basal 2 mm connate into a cup, apices acuminate, glabrous; petals 5, stamens 19, ovaries glabrous, style apices shallowly 3-cleft.

Guangxi: Quan Xian, Shanchuan Village, Baoding Shan, Zhong Jixin (T. S. Tsoong) 81616 (holotype in SCBI, isotype in A).

Sealy misidentified this specimen as *C. acutisepala*, but in reality it is a very unusual new species with the tip of the sepals 9 mm long. *C. acutisepala* described by Tsai and Feng is, however, only a variety of *C. forrestii* with pointed sepals.

6. ***Camellia forrestii*** (Diels) Cohen-Stuart, Meded. Proefst. Thee 40: 68. 1916.
   Thea forrestii Diels, Notes Roy. Bot. Gard. Edinb. 5: 284. 1912.
   Thea polygama Hu, Bull. Fan Mem. Inst. Biol. Bot 8: 132. 1935.
   Theopsis forrestii (Diels) Nakai, Journ. Jap. Bot. 16: 705. 1940.
   Theopsis polygama (Hu) Nakai, Journ Jap. Bot. 16: 706. 1940.
   Camellia liuii Tsai & Feng, Acta Phytotax. 1: 189. 1951.

6a. ***Camellia forrestii*** (Diels) Cohen-Stuart var. ***forrestii***

Branchlets pubescent. Leaves short and small, elliptic, 2.2-3 cm long, 1.5-3 cm wide. Pedicels short, bracts 3-4, sepals ciliate, 3-4 cm long, ovaries pubescent, stamens without a filament tube.

Yunnan: Yimen, Yin Wenqing 79, 239; Shuangbai, elev. 2250 m, Liu Weixin 429; Yun Xian, Zhu Taiping 539; Fengqing, Zhu Taiping 560; Wei Shan, Jiang Ying (Y. Tsiang) 12021, 13104; Mengzi, Liu Shenno (S. N. Liu) 16618, 16594; Jingdong, Feng Guomei (K. M. Feng) 1951; Kunming Heilongtan, Zhang Hongda (H. T. Chang) 5121.

Common in Yunnan, seeds can be used for oil.

6b. ***Camellia forrestii*** var. ***acutisepala*** (Tsai & Feng) Chang, Acta Sci. Nat. Univ. Sunyatseni, monogr. ser. 1: 138. 1981.
   Camellia acutisepala Tsai & Feng, Acta Phytotax. 1: 189. 1951.

This variety is very close to *C. forrestii* var. *forrestii* only the sepals are triangular-ovate, apices acute.

Yunnan: Jingdong, Huangcaoling, Li Minggang 1951 (holotype in PE); same loc., Yang Zenghong 101390, 101493; same loc., Qiu Bingyun 53266, 53459, 53463; Fumin, Qiu Bingyun 58763.

Sealy in his Rev. Gen. Camellia considered Zhong 81616 from Guangxi to belong to this species, but it belongs to the species *C. longicalyx*.

Li 1951 differs from *C. forrestii* var. *forrestii* only by the sepals being pointed.

7. ***Camellia buxifolia*** Chang, Acta Sci. Nat. Univ. Sunyatseni, monogr. ser. 1: 139. 1981.

Species *C. rosthornianae* proxima, sed foliis brevioribus 2-3 cm longis, staminibus subliberis, fructibus ellipsoideis, 2-3-locularibus, valvis crassioribus 1 mm crassis distincta.

Frutex 1.5-3 m altus, ramulis sparse pubescentibus. Folia coriacea ovata vel elliptica 2-3 cm longa 1-1.6 cm lata, apice acuta, acumene obtuso, basi late cuneata, supra nitidula ad costam sparse puberula, subtus ad basi costae puberula excepta glabra, nervis lateralibus circ. 5-jugis subtus inconspicuis supra leviter impressis, margine serrulata, petiolis 1-1.5 mm longis sparse puberulis. Flores terminales vel axillares, pedicellis 2 mm longis; bracteis 4 ovatis 1-1.5 mm longis ciliatis; sepalis 5 late ovatis 2 mm longis apice rotundatis ciliatis; petalis albi 1 cm longis glabris, exterioribus 2 oribuclatis 7-8 mm longis basi leviter connatis, interioribus 3 ovoideis circ. 1 cm longis;

Camellia forrestii (Diels) Cohen-Stuart
1. fruiting branch; 2. flower.

54

staminibus 8-9 mm longis liberis basi ad petala adnatis; ovariis glabris, stylis 7-8 mm longis apice 3-fidis. Capsula pyriformis 1 cm longis 7-8 mm diam. 2-3-locularis, pericarpio 1 mm crasso; semina singula; pedicellis dilatatis 5-6 mm longis.

Shrub 1.5-3 m tall, branchlets sparsely pubescent. Leaves coriaceous, ovate or elliptic, 2-3 cm long, 1-1.6 cm wide, apices acute or obtuse, bases broadly cuneate; dark green above, slightly shiny in the dry state, glabrous or base of midvein slightly puberulent; glabrous below except slightly pubescent along the base of the midvein; lateral veins ca. 5 pairs, impressed above, inconspicuous below; margins sparsely serrate, teeth separation 2-3 mm; petioles 1-1.5 mm long, slightly puberulent. Flowers terminal or axillary, white, 1 cm long, glabrous; pedicels 2 mm long; bracts 4, ovate, 1-1.5 mm long, ciliate; sepals 5, broadly ovate, 2 mm long, apices obtuse, ciliate; petals 5, outer 2 suborbicular, 7-8 mm long, bases slightly connate, inner 3 basally adnate with stamens for ca. 2 mm, obovoid, 1 cm long; stamens 8-9 mm long, outer stamens adnate with petals, others free, glabrous; ovaries glabrous; styles 7-8 mm long, glabrous, apices shallowly 3-cleft. Capsules pyriform, 1 cm long, 7-8 mm wide, 2-3-locular, 1 seed per locule, valves ca 1 mm thick, base with persistent sepals and bracts; pedicels dilated, 5-6 mm long.

Sichuan: Nanchuan, Jinfo Shan, Daheba, Li Guofeng 64098, 64923, 65062, 64889; Nanchuan, Luoguotang, Xiong Jihua and Zhou Zilin 91737; Emei Shan, Yang Guanghui 52895, 53121 (type in SYS); Chengdu, Caotang, cultivated, Zhang Hongda (H. T. Chang) 5509.

Hubei: Xingshan, elev. 200 m, Li Hongjun 1861.

8. *Camellia minutiflora* Chang, Acta Sci. Nat. Univ. Sunyatseni, monogr. ser. 1: 140. 1981.

Species C. *euryoidi* assimilis, sed floribus minoribus, staminibus subliberis, pedicellis multo brevioribus differt.

Frutex, ramulis puberulis in sicco atro-brunneis; foliis anguste oblongis vel lanceolatis 2-3.5 cm longis 6-9 mm latis, apice subacutis basi cuneatis, utrinque glabris, margine crenatis, nervis lateralibus circ. 6-jugis obscuris vel in sicco supra impressis, petiolis 1-2 mm longis. Flores albi 1-2 axillares parvi, pedicellis 1-2 mm longis, bracteis 4-5 glabris 0.3-0.7 mm longis; sepalis 5 lanceolatis 1-1.5 mm longis glabris; petalis 5-6 obovatis 6-8 mm longis 4-6 mm latis mucronatis glabris; staminibus 5-7 mm longis glabris subliberis vel basi 1 mm connatis, antheris basifixis; ovariis glabris, stylis 5-7 mm longis apice brevius 3-fidis.

Shrub; branchlets puberulent, dull brown when dry. Leaves narrowly oblong or lanceolate, 2-3.5 cm long, 6-9 mm wide, apices subacute, bases cuneate, both surfaces pubescent, margins serrate; lateral veins ca. 6 pairs, obscure or impressed above in the dry state; petioles 1-2 mm long. Flowers white, 1-2 axillary, slender and small, pedicels 1-2 mm long; bracts 4-5, 0.3-0.7 mm long, glabrous; sepals 5, lanceolate, 1-1.5 mm long, glabrous; petals 5-6, obovate, 6-8 mm long, 4-6 mm wide, apices mucronate, glabrous; stamens 5-7 mm long, glabrous, nearly free or slightly connate, anthers basifixed; ovaries glabrous, styles 5-7 mm long, apices briefly 3-cleft.

Hong Kong: Mt. Ba-sin-ling, 12 August 1964, Chinese Univ. Hong Kong Herb. 67 (135) (holotype in SYS no. 146509).

9. *Camellia parvicuspidata* Chang, Acta Sci. Nat. Univ. Sunyatseni, monogr. ser. 1: 141. 1981.

Species C. *cuspidatae* proxima, a qua ramulis pubescentibus, foliis anguste lanceolatis, pedicellis brevioribus, sepalis subliberis minoribus ovatis, petalis staminibusque multo brevibus differt.

Frutex, ramulis, puberulis. Folia coriacea anguste lanceolata 4-6 cm longa 1-1.5 cm lata, apice caudato-acuminata, caudo 1-1.5 cm longo, basi late cuneata, supra nitidula ad costam puberula, subtus glabra, nervis lateralibus circ. 6-jugis utrinque inconspicuis, margine serrulata, petiolis 1-2 mm longis puberulis. Flores terminales, pedicellis 1-2 mm longis; bracteis 5 ovatis 1-2 mm longis acutis glabris; sepalis 5 ovatis 4 mm longis apice acutis glabris; petalis 7 albis obovatis 8-11 mm longis basi connatis glabris; staminibus 7-9 mm longis liberis vel leviter connatis glabris; ovariis glabris, stylis 8-12 mm longis, apice 3-fidis.

Shrub, branchlets puberulent. Leaves coriaceous, narrowly lanceolate, 4-6 cm long, 1-1.5 cm wide, apices caudate-acuminate with a 1-1.5 cm long acumen, bases broadly cuneate; greyish-brown above in the dry state, not shiny, midveins puberulent; glabrous below; lateral veins ca. 6 pairs, obscure on both surfaces; margins serrulate, teeth separation 1-3 mm; petioles 1-2 mm long, puberulent. Flowers terminal, pedicels 1-2 mm long; bracts 5, ovate, apices acute, 1-2 mm long, glabrous; sepals 5, free or bases slightly connate, ovate, apices acute, 4 mm long, glabrous; petals 7, white, basally connate, obovate, 8-11 mm long, glabrous; stamens 7-9 mm long, filaments free or bases slightly connate, glabrous; ovaries glabrous; styles 8-12 mm long, apices shallowly 3-cleft. Capsules ovate, 1 cm long, 3-locular. Flowering period March to April.

Guangdong: Beijiang Yao Shan, Xin Shuzhi 9112 (holotype in SYS); Chaoan, Fenghuang Shan, 18 November 1959, SCBI Plant Geog. Dept. 08; Jiaoling, Shihu, Deng Liang 4654, 4851; Fengshun, Beixi, Li Xuegen 200988; Ruyuan, Deng Liang 5834.

10. *Camellia acutissima* Chang, Acta Sci. Nat. Univ. Sunyatseni, monogr. ser. 1: 142. 1981.

A *C. cuspidata* differt ramulis puberulis, foliis minoribus pedicellis brevioribus, sepalis minoribus.

Frutex, ramulis puberulis. Folia coriacea ovato-lanceolata 4-5.5 cm longa 1-1.3 cm lata, apice caudata, caudo 9-13 mm longo, basi late cuneata, supra nitidula ad costam puberula subtus glabra, nervis lateralibus 5-jugis inconspicuis, margine dense serrulata, petiolis 3 mm longis pubescentibus. Flores ignoti. Capsula 1-3 terminalis vel solitaria axillaris, pedicellis 3 mm longis; bracteis 5 semiorbiculatis circ. 1 mm longis ad pedicellum dispersis; calice 5 mm longo inferiore connato calatho, lobis 5 semiorbiculatis 2.5 mm longis apice rotundatis ciliatis; ovariis glabris, stylis apice 3-fidis. Capsula (immatura) glabra, pericarpio 1.5 mm crasso, 3-locularis, semina singula in quoque loculo.

Shrub, branchlets brown puberulent; old branches glabrous, brown. Leaves coriaceous, ovate-lanceolate, 4-5.5 cm long, 1-1.3 cm wide, apices caudate with a 9-13 mm long cauda, bases broadly cuneate or obtuse; dark green above in the dry state, slightly shiny, midvein puberulent; brownish-green below, glabrous; lateral veins ca. 5 pairs, very obscure; margins densely serrate, teeth separation 1-2 mm; petioles 3 mm long, pubescent. Capsules 1-3 terminal, or solitary axillary; pedicels 5 mm long, glabrous, upper end thickened; bracts 5, semiorbicular, 1 mm long, born on the pedicel, glabrous; sepals 5 mm long, lower half connate cup-shaped, margins membranous, easily broken, exterior usually glabrous, 5-cleft, clefts semiorbicular, 2-2.5 mm long, apices rounded, ciliate; ovaries glabrous. Capsules ovoid, 2 cm long, valves 2 mm thick, 1-locular, 1-2 seeds per locule.

Guangdong: Ruyuan, Xinqiao Shan, mountainous valley in dense forests, Liu Yingguang 355 (holotype in SCBI); Luofu Shan, Luofuding, Yan Suyzhou s.n. (SYS no. 141788).

Hunan: Nanyue, in the forests behind Shangfeng temple, elev. 1260 m, flowers white, Liu Linhan (H. L. Liu) 15716.

This species is similar to *C. cuspidata* only the branchlets, leaf midveins, petioles and sepals are puberulent; sepals connate or cup-shaped; leaf blades thin and small, ovate-lanceolate, apices caudate; pedicels shorter, sepals smaller.

11. *Camellia subacutissima* Chang, Acta Sci. Nat. Univ. Sunyatseni, monogr. ser. 1: 143. 1981.

Species *C. acutissimae* affinis, sed foliis majoribus 6 cm longis 2 cm latis, pedicellis brevissimis, sepalis minoribus 2-3 mm longis distincta.

Frutex, ramulis brunneo-pubescentibus. Folia coriacea ovato-lanceolata 5-6 cm longa 1.5-2 cm lata, apice caudato-acuminata, basi rotundata vel late cuneata, supra in sicco nitida ad costam puberula, subtus primo pubescentia demum glabrescentia vel ad costam puberula, nervis lateralibus 5-6-jugis utrinque inconspicuis, margine serrulata, petiolis 4-5 mm longis pubescentibus. Flores non visi. Capsula axillaris globosa 1-1.5 cm in diametro glabra 1-locularis 3-valvata dehiscens; semina solitaria; pedicellis brevissimis; bracteis persistenibus 4 semiorbiculatis 1 mm longis; sepalis persistentibus late ovatis 2.5-3 mm longis glabris.

55 *Camellia acutissima* Chang fruiting branch.

Shrub, branchlets with brown pubescence. Leaves coriaceous, ovate-lanceolate, 5-6 cm long, 1.5-2 cm wide, apices caudate-acuminate, bases rounded or obtuse; shiny above in the dry state, pubescent along the midvein; at first pubescent below, later glabrous or midvein pubescent; lateral veins 5-6 pairs, more or less obscure on both surfaces; margins serrulate; petioles 4-5 mm long, abundantly pubescent. Flowers not seen. Capsules axillary, globose, 1-1.5 cm in diameter, pubescent, 1-locular, 1-seeded, 3-valvate dehiscent, pedicels extremely short; persistent bracts 4, semiorbicular, 1 mm long, glabrous; sepals broadly ovate, 2.5-3 mm long, glabrous.

Guangxi: Wudidian, Liang Shengye (S. Y. Liang) 1327, 1377 (holotype in SYS); Qin Renchang (R. C. Ching) 6098; Damiao Shan, Chen Shaoqing (S. C. Chun) 16681.

Hunan: Qianyang, Erqu, Taipo Shan, Li Zetang 2996.

This species is close to *C. acutissima* only the pedicels are shorter; sepals smaller, 2-3 mm long, nearly free, basally not connate into a cup.

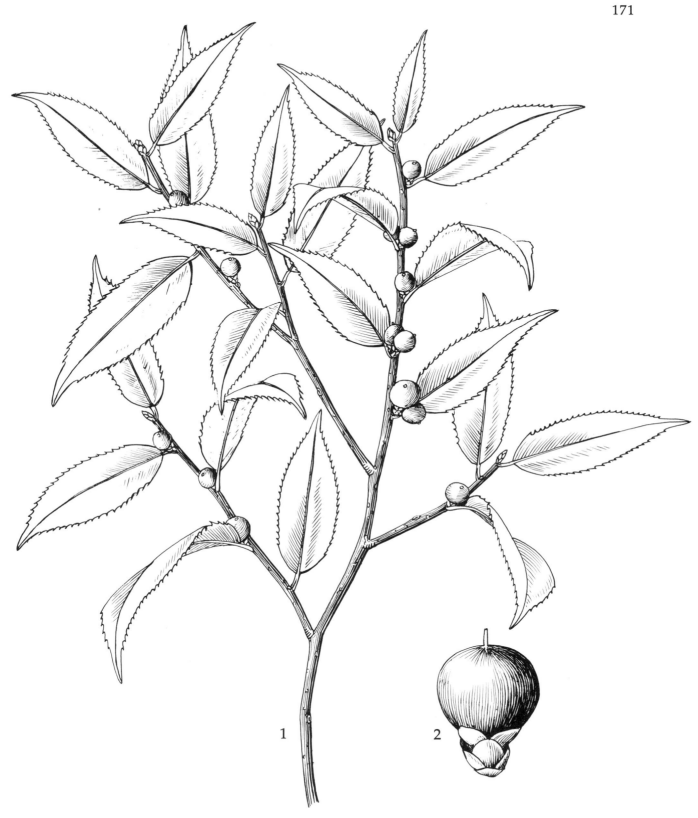

Camellia subacutissima Chang
1. fruiting branch; 2. capsule.

12. **Camellia callidonta** Chang, Acta Sci. Nat. Univ. Sunyatseni, monogr. ser. 1: 143. 1981.

A *C. subacutissima* differt foliis membranaceis majoribus, nervis lateralibus pluribus circ. 8-10 utrinsecus, margine acriter serratis, pedicellis longioribus.

Arbor parvi circ. 3-6 m alta, ramulis pubescentibus. Folia tenuia submembranacea lanceolata 5-7.5 cm longa 1.5-2 cm lata, apice acuminata vel caudata, basi late cuneata, supra opaca ad costam puberula, basi glabra, nervis lateralibus 8-10-jugis, margine acriter serrata, petiolis 2-4 mm longis. Flores ignota. Capsula solitaria globosa 1.3 cm in diametro glabra 1-locularis 3-valvata dehiscens, valvis tenuibus circ. 0.6 mm crassis; semina solitaria; bracteis persistentibus 4 parvis; sepalis 5 circ. 2 mm longis glabris; pedicellis 4-5 mm longis.

Small tree 3-6 m tall, branchlets pubescent. Leaves thinly membranous, lanceolate, 5-7.5 cm long, 1.5-2 cm wide, apices acuminate or caudate, bases broadly cuneate; not shiny above, mid-region pubescent; glabrous below; lateral veins 8-10 pairs, very fine; margins sharply serrate, petioles 2-4 mm long. Capsules solitary, globose, 1.3 cm in diameter, glabrous, 1-locular, 1 seeded, 3-valvate dehiscent, valves 0.6 mm thick, seeds globose; persistent bracts 4, narrow and small; sepals 5, 2 mm long, glabrous; pedicels 4-5 mm long, glabrous.

Yunnan: Yingjiang, Kunaqian Shan, elev. 1800 m, Yang Zenghong 6916 (holotype in KUN); same loc., Tongbi Commune, elev. 1350 m, Tao Guoda 13359.

This species is relatively close to *C. subacutissima* only the leaf blades are thinly membranous, leaves longer, margins sharply serrate, and lateral veins 8-10 pairs.

SER. II.
Gymnandrae Chang, Acta Sci. Nat. Univ. Sunyatseni, monogr. ser. 1: 144. 1981.

    Filamenta basi connata tubifera glabra, ovaria glabra.
    Filaments connate into a short tube, glabrous; ovaries glabrous.
    24 species; 23 in China south of the Changjiang (Yangtse River).
    Type: *C. fraterna* Hance

13. **Camellia handelii** Sealy, Kew Bull. 1949(2): 219. 1949.
Thea fraterna Hand.-Mazz., Symb. Sin. 7: 393. 1931, *non* Hance.

Branchlets velutinous. Leaves ovate-elliptic, 3-4 cm long. Sepals long-sericeous, stamens with a short filament tube.

Guizhou: Zunyi, large tree, Zhong Buqin (P. C. Tsoong).

Hunan: Zhangsha, Yuelu Shan, November 1963, Zhang Hongda (H. T. Chang) 5016 (type loc.); Dongkou Xian, Liu Linhan (H. L. Liu) and He Guanzhou 16602; Yuelu Shan, Li Binggui 168; Wugang, Liu Linhan (H. L. Liu) and He Guanzhou 16257, 15573; Yuelu Shan, H. T. Chow 45336.

Jiangxi: Jian, 263 Group 434.

Distribution: Hunan, Jiangxi, Guizhou.

This species is close to *C. forrestii* only the pedicels are shorter, sepals ciliate, filaments free.

14. **Camellia triantha** Chang, Acta Sci. Nat. Univ. Sunyatseni, monogr. ser. 1: 144. 1981.

Species *C. caudatae* subsimilis, sed filamentis glabris, ovariis stylisque omnino glaberrimis differt; in sectione *Theopsi* illa *C. stuartianae* similis, quae differt tubis filamentorum brevioribus, carpellis 5, stylis 5-fidis.

Frutex, ramulis puberulis. Folia membranacea elliptica vel obovato-oblonga 8-10.5 cm longa 3-4 cm lata, apice caudata, caudo 1-1.5 cm longo, basi obtusa, supra opaca ad

Camellia handelii Sealy
1. fruiting branch; 2. capsule.

Camellia costei Lév.
1. fruiting branch; 2. capsule.

costam puberula, subtus ad costam pilosa in petiolis 3-5 mm longis pubescentibus. Flores terni terminales vel singuli axillares albi, pedicellis 3-4 mm longis; bracteis 5 semilunatis vel orbiculatis 1-1.5 mm longis pubescentibus; sepalis 5 orbiculatis 2.5-3 mm longis basi leviter connatis griseo-pubescentibus; petalis 5, exterioribus 2 late ovatis pubescentibus, interioribus 3 obovatis basi connatis pubescentibus; staminibus petalis brevioribus, basi ad petala adnatis, extimis connatis, tubo filamentorum ⅔ quamdiu staminibus, glabris; ovariis glabris, stylis staminibus brevioribus, apice 3-fidis.

Shrub, branchlets puberulent. Leaves membranous, elliptic or ovate-oblong, 8-10.5 cm long, 3-4 cm wide, apices narrowly caudate-acuminate with a 1-1.5 cm long cauda, bases broadly cuneate or obtuse; brownish-green above in the dry state, not shiny, midveins puberulent; gray below, pilose along the midvein, surface somewhat tuberculate; lateral veins 7-8 pairs, conspicuous above or slightly impressed, slightly protruding below; margins densely serrulate, teeth separation 1-1.5 mm; petioles 3-5 mm long, pubescent. Flowers 3 terminal or 1 axillary; pedicels 3-4 mm long, with greyish pubescence; bracts 5, semiorbicular or orbicular, 1-1.5 mm long, greyish pubescent; sepals 5, bases slightly connate, orbicular, 2.5-3 mm long, brown pubescent; petals 5, outer two broadly ovate, exteriors greyish pubescent, inner 3 obovate, exteriors pubescent, bases slightly connate; stamens shorter than the petals, bases slightly adnate with the petals, outer filament whorl connate into a tube for ⅔ the filament length, free filaments short, glabrous; ovaries glabrous; styles shorter than the stamens, glabrous, apices shallowly 3-cleft.

Guangxi: Yongding, Huang Fengsheng 17625 (holotype in SYS).

15. *Camellia costei* Lév. in Fedde, Repert. Sp. Nov. 10: 148. 1911.
 Thea costei (Lév.) Rehd., Journ. Arn. Arb. 15: 98. 1934.
 Hubei: Chengfeng, Li Hongjun 6973.
 Guizhou: Pingzhou, near the Guangxi border, Jiang Ying (Y. Tsiang) 7104, 8008.
 Hunan: Xuefeng Shan, Li Zetang 1049; Dongkou, Yuexi, Li Zetang 2634.
 Jiangxi: Nie Minxiang 2622.
 Guangxi: Qin Renchang (R. C. Ching) 6969; Xiang Xian, Huang Zhi (C. Wang) 39474.
 Guangdong: Luoding, Yadao Dashan, Wang Bosun 171, 466; Yingde, Xie Xi 453.
 Yunnan: Zhengxiong, Yang Jingsheng and Liu Dachang 2862; Guangnan, Wu Qiannan 9814; Guangnan, Lin Wenzhong 619; Guangnan, Wang Qiwu (C. W. Wang) 87468.

This species is very close to *C. handelii* only the leaf blades are larger; pedicels longer, 4-5 mm long; flowers pubescent, shorter.

16. *Camellia tsaii* Hu, Bull. Fan Mem. Inst. Biol. Bot. 8: 132. 1938.
 Thea tsaii (Hu) Gagnep., Suppl. Fl. Gén. Indo-Chine 1: 316. 1943.
 Thea fusiger Gagnep., Not. Syst. 10: 126. 1942.

16a. *Camellia tsaii* Hu var. *tsaii*
Branchlets pubescent. Leaves oblong 6-9.5 cm long, to 2.9 cm wide. Pedicels 4-5 mm long; sepals pubescent, 1.5-3 mm long; filaments connate into a tube, and ovaries glabrous.
 Yunnan: Cai Xitao (H. T. Tsai) 56363.
 Vietnam: M. Poilane 19040.
 Distribution: Yunnan, Burma and northern Vietnam.

1-7. *Camellia tsaii* Hu 8-9. *Camellia trichoclada* (Rehd.) Chien
1. flowering branch; 2. ovary longitudinal section; 3. ovary cross section;
4. calyx and styles; 5. petal and stamens; 6. capsule and seed; 7. leaf detail;
8. fruiting branch; 9. capsule.

Camellia tsaii Hu var. *synaptica* (Sealy) Chang
1. flowering branch; 2. flower; 3. calyx and pistil.

16b. ***Camellia tsaii*** var. ***synaptica*** (Sealy) Chang, Acta Sci. Nat. Univ. Sunyatseni, monogr. ser. 1: 148. 1981.

Camellia synaptica Sealy, Kew Bull. 1949(2): 221. 1949.

Yunnan: Sino-Soviet Expedition 3042, 1440; Zhaotong, Li Xiwen 241.

Sichuan: Emei Shan, Yang Guanghui 49913, 52065, 52146, 53289, 53373, 53415, 53498, 53603, 53625, 53692, 53771; same loc., Zhang Hongda (H. T. Chang) 5542.

Hunan: Xuefeng Shan, Li Zetang 1657.

The difference between this variety and the original species is that the young shoots are thicker and they do not droop, and the leaves thick and shinier. The basic characteristics of the flowers are the same except the pedicels are longer and the back of the bracts are glabrous.

17. ***Camellia transnokoensis*** Hay., Icon. Pl. Formos. 8: 11. 1919.

Thea transnokoensis (Hay,) Mak. & Nem., Fl. Jap., ed. 2, 745. 1931.

Theopsis transnokoensis (Hay.) Nakai, Journ. Jap. Bot. 16: 707. 1940.

Branchlets pubescent. Leaves oblong, 3-5 cm long. Pedicels 2.5 mm long; sepals 2-3 mm long, free, only ciliate; stamens forming a short filament tube, ovaries glabrous.

Taiwan: Kanehira et Saski 48.

18. ***Camellia rosthorniana*** Hand.-Mazz., Anz. Akad. Wiss. Math. Nat. Mien 61: 108. 1925.

Thea rosthorniana (Hand.-Mazz.) Hand.-Mazz., Symb. Sin. 7: 394. 1931.

Yunnan: Cai Xitao (H. T. Tsai) 51796.

Sichuan: Emei Shan, Yang Guanghui 53466; Qu Guiling 6806; Nanchuan, Li Guofeng 6042.

Guangxi: Longsheng, Tan Haofu and Li Zhongdi 71057.

Guizhou: Luodian, Zhang Zhisong and Zhang Yongtian 246.

Distribution: Sichuan, Hunan, Hubei, Guangxi, Guizhou and Yunnan.

This species is very close to *C. transnokoensis* only the outer petals are free and pedicels shorter. Only the Sichuan specimens have connate sepals and pedicels to 4 mm long. With the above Guizhou and Guangxi specimens the leaf blades are lanceolate with caudate apices.

19. ***Camellia lutchuensis*** Ito in Ito and Matsumura, Tent. Fl. Luch., 332. 1909.

Thea lutchuensis (Ito) Hay., Mat. Fl. Form., 45. 1911.

Theopsis lutchuensis (Ito) Nakai, Journ. Jap. Bot. 16: 706. 1949.

Branchlets pubescent. Leaves oblong, 2.5-4 cm long, 1-1.8 cm wide. Pedicels 2.5 mm long, bracts 8; sepals 2 mm long, pubescent; stamens with a short filament tube.

Japan: Onna, Okinawa, 15 January 1924, Tsiro s.n.; Taiwan For. Inst. 27801.

20. ***Camellia euryoides*** Lindl., Bot. Reg. 12: sub. pl. 983. 1826.

Thea euryoides (Lindl.) Booth, Trans. Hort. Soc. London 7: 560. 1830.

Theaphylla euryoides (Lindl.) Raf., Sylva Tellur., 139. 1838.

Camellia theiformis Hance, Ann. Sci. Nat. Paris, ser. 4, 15: 221. 1861.

Thea theiformis (Hance) O. Kuntze, Rev. Gen. Pl., 64. 1891.

Branchlets pubescent. Leaves small, oblong, 2-4 cm long, both ends acute, back pubescent. Pedicels 6-9 mm long, glabrous; sepals glabrous, filament tube short, ovaries glabrous. Capsules 7-9 mm wide, 1-locular.

Guangdong: Xingning, SCBI Plant Geog. Dept. 7309; Dapu, Zeng Huaide (W. T. Tsang) 21078; Renhua, Zeng Huaide (W. T. Tsang) 26090, 31198.

Distribution: Fujian, Guangdong and Jiangxi.

Camellia rosthorniana Hand.-Mazz.
1. flowering branch; 2. calyx and stamens; 3. calyx and style.

21. ***Camellia trichoclada*** (Rehd.) Chien, Contr. Biol. Lab. Sci. Soc. China Bot. 12: 100. 1939.

Thea trichoclada Rehd. in Rehd. and Wils., Journ. Arn. Arb. 8: 176. 1927.
Theopsis trichoclada (Rehd.) Nakai, Journ. Jap. Bot. 16: 707. 1940.

Branchlets with long pubescence. Leaves small, elliptic 1.5-2 cm long, apices obtuse, bases rounded. Pedicels 3-4 mm long; sepals 1-2 mm long, glabrous. Fruit glabrous.

Fujian: Yongtai, Dunqianlun, Lin Yong 5490.

Zhejiang: Y. L. Keng 324 (holotype in A); Pingyang, HZBG 5898, 7412; Taishun, Wang Jingxiang 1303.

22. *Camellia parvilimba* Merr. & Metc., Lingnan Sci. Journ. 16: 171. 1937.
Theopsis parvilimba (Merr. & Metc.) Nakai, Journ. Jap. Bot. 16: 707. 1940.

22a. *Camellia parvilimba* Merr. & Metc. var. *parvilimba*

This species is very close to *C. euryoides* only the branches are very divided, branchlets more slender; leaves shorter, glabrous above; pedicels slightly longer (1 cm); sepals 5, 1.5-2 mm long, glabrous; petals 6, 1.5 cm long, broadly ovate; stamens and petals equal in length, filament tube 5-6 mm long, ovaries glabrous, capsules 1 cm in diameter.

Guangdong: Dapu, Zeng Huaide (W. T. Tseng) 21560 (holotype in SCBI, isotype in A); Pingyuan, Deng Liang 4357; Mei Xian, Li Xuegen 201116; Fengshun, Li Xuegen

Camellia parvilimba Merr. & Metc.
1. flowering branch; 2. leaf; 3. capsule; 4. seed.

200924, 201116; Renhua, Deng Liang 7353; Mei Xian, SCBI Plant Geog. Dept. 7351.

Jiangxi: Huichang, Yifeng Commune, Zengkeng, Yang Xianxue 12771; Gannan, Dayaokeng, J. L. Gressitt 1601.

22b. *Camellia parvilimba* var. *brevipes* Chang, Acta Sci. Nat. Univ. Sunyatseni, monogr. ser. 1: 151. 1981.

A typo pedicellis brevioribus circ. 4 mm longis differt.

Small tree, branchlets pubescent. Leaves ovate, 2-3 cm long, 8-13 mm wide, apices acuminate, bases slightly rounded. Fruit pedicels 4 mm long; bracts 4-5, shattering; sepals broadly ovate, 2 mm long, glabrous. Capsules 1 cm wide.

Jiangxi: Wugong Shan, Lai Shushen 1699 (holotype in HLG); Yongxin, Lian Shan Kenzhichang, Lai Shushen 949; Suichuan, Dafen, Yue Housan 4478.

23. *Camellia elongata* (Rehd. & Wils.) Rehd., Journ. Arn. Arb. 3: 224. 1922.

Thea elongata Rehd. & Wils. in Sargent, Pl. Wils. 2: 392. 1915.
Thea caudata var. *faberi* Kochs in Engler, Bot. Jahrb. 27: 583. 1900.
Theopsis elongata (Rehd. & Wils.) Nakai, Journ. Jap. Bot. 16: 705. 1940.

Branchlets glabrous. Leaves narrowly lanceolate, apices long-pointed, glabrous. Pedicels 1 cm long; calyx cup-shaped, glabrous; outer filament whorl nearly totally connate into a tube, inner free filaments whorl pubescent. Capsules ovate.

Sichuan: Emei Shan, Fang Wenpei (W. P. Fang) 7797; same loc., Yang Guanghui 52703, 53270, 55985, 56029, 57518; same loc., Zhang Hongda (H. T. Chang) 5628, 5629, 5638.

Guizhou: Wudidian, Guizhou Agr. Inst. 11.

24. *Camellia longicarpa* Chang ex Chang, *sp. nov.*; Chang, Acta Sci. Nat. Univ. Sunyatseni, monogr. ser. 1: 154. 1981, *nom. invalid.*

Species foliis majoribus, pedicellis robustis, sepalis 5-6 mm longis, basi connatis, ovariis ovoideis apice acutis, capsulis ellipoideis distincta.

Frutex circ. 3 m altus, ramulis primo griseo-pilosis mox glabrescentibus. Folia coriacea elliptica 6-8.5 cm longa 2-3 cm lata, apice acuminata vel caudata, caudo 1-1.5 cm longo, basi late cuneata, supra opaca ad costam puberula, subtus ad costam pilosa, nervis lateralibus 7-9-jugis utrinque visibilibus, margine superiore serrulata, petiolis 3-4 mm longis hirtellis. Flores singuli terminales albi; pedicellis 5 mm longis robustis; bracteis 5-6, inferioribus 3 ad medium pedicelli dispersis, superioribus 2-3 sub sepalis adnatis; sepalis 5 basi connatis calathis 5-6 mm longis glabris, limbo suborbiculato, margine scarioso; petalis 6-7, 1.5-2 cm longis, exterioribus 2 pubescentibus; staminibus petala subaequalibus, extimis connatis tubulatis; ovariis glabris, stylis apice 3-fidis. Capsula ellipsoidea 2.5 cm longa 1.5 cm lata, 1-3-locularis, pericarpio lignoso 1.5-2 mm crasso; semina singula in quoque loculo.

Tree 3 m tall; branchlets at first greyish-white pilose, soon glabrous. Leaves coriaceous, elliptic, 6-8.5 cm long, 2-3 cm wide, apices acuminate or caudate with a 1-1.5 cm long cauda, bases broadly cuneate or obtuse; greyish-brown above in the dry state, not shiny, midvein brown-puberulent; base of the midvein pilose below; lateral veins 7-9 pairs, equally conspicuous on both surfaces; upper half of margins serrulate; petioles 3-4 mm long, hirtellous. Flowers terminal; pedicels 5 mm long, thick; bracts 5-6, 3 on the middle of the pedicel, crescent-shaped, glabrous, margins thinly membranous, 1-2.5 mm long; 2-3 bracts relatively large, born near the sepals; calyx shallowly cup-shaped, 5-6 mm long, glabrous, sepals semiorbicular, margins thin; petals white, 6-7, outer 2 somewhat brownish-pubescent, 1.5 cm long, the remaining petals 2 cm long, glabrous; stamen length ca. equal that of petals, outer filament whorl basally connate into a tube; free filaments shorter, glabrous; ovaries glabrous; style length equal to stamens, shallowly 3-cleft. Capsules elliptic-ovate, 2.5 cm long, 1.5 cm wide, 3-locular or 1-locular, 1 seed per locule; valves woody, 1.5-2 mm thick.

Sichuan: Emei Shan, Bailudong, Xiao Yongxian 48217 (*holotypus* in SCBI); Emei Shan, Yang Guanghui 53898, 56192.

This species is close to *C. tsaii* var. *synaptica* only the calyx is much longer and not ciliate, and the fruit is longer. The species differs from *C. cuspidata* by having pubescent branches, filaments connate into a short tube and fruit longer.

63 *Camellia elongata* (Rehd. & Wils.) Rehd.
　1. flowering branch; 2. petals and stamens; 3. calyx and style;
　4. capsule.

1. *Camellia cratera* Chang 2-3. *Camellia longicarpa* Chang ex Chang
1. flowering and fruiting branch; 2. flowering branch; 3. fruiting branch.

25. *Camellia parvilapidea* Chang, Acta Sci. Nat. Univ. Sunyatseni, monogr. ser. 1: 155. 1981.

A *C. stuartiana* differt sepalis majoribus, ovariis 3-locularibus, capsulis minoribus, pericarpio multo crassiore; a *C. assimili* sepalis majoribus, ovariis glabris, pericarpio crassiore recedit.

Arbor parva circ. 7-8 m alta, ramulis puberulis. Folia coriacea vel tenuiter coriacea, elliptica vel oblonga 6-10 cm longa 2.5-3 cm lata, apice caudata, caudo 1.5-2 cm longo, basi cuneata, utrinque ad costam puberula muricatula, nervis lateralibus 10-11-jugis, margine serrulata, petiolis 4-5 mm longis pubescentibus. Flores non visi. Capsula globosa 1.6 cm in diametro 3-locularis, pericarpio 3 mm crasso, glabro; semina solitaria in quoque loculo; pedicellis 3-4 mm longis robustis; bracteis persistentibus 5 semilunatis vel suborbiculatis 1-3 mm longis pubescentibus; sepalis basi connatis calathis pubescentibus, limbis 5 semiorbicularibus 5-6 mm longis.

Small tree 7-8 m tall, branchlets puberulous; older branches glabrous, brown. Leaves coriaceous or thinly coriaceous, elliptic or oblong, 6-10 cm long, 2.5-3 cm wide, apices caudate, cauda 1.5-2 cm long, bases cuneate or broadly cuneate; light brown above in the dry state, not shiny, with numerous small tubercules, midvein pubescent; lateral veins 10-11 pairs, evident above, slightly protruding below; margins sharply serrulate; petioles 4-5 mm long, pubescent. Flowers not seen. Fruit axillary or terminal; pedicels 3-4 mm long, thick; bracts 5, semiorbicular or suborbicular, 1-3 mm long, pubescent; calyx shallowly cup-shaped, 5-6 mm long, pubescent, basal half connate, upper half divided into 5 sepals, 2-3 mm long. Capsules globose, 1.6 cm in diameter, 3-locular, 1 seed per locule; pericarp 3 mm thick, glabrous.

Guangdong: Shantou, Dayang Shan, F. A. McClure 6504, 6619 (holotype in SYS).

Guangxi: Damiao Shan, Jiuwan Shan, near Shuanghe, Chen Shaoqing (S. C. Chen) 14370, 14382, 14386, 14658, 15005, 15746, 16006; same loc., Chen Dezhao 788; Qin Renchang (R. C. Ching) 5818.

26. *Camellia stuartiana* Sealy, Kew Bull. 1949(2): 220. 1949.

Small tree, branchlets pubescent. Leaves elliptic, 8-12 cm long, 3-4.5 cm wide, caudate, pubescent above. Pedicels 3 mm long, bracts 5; calyx cup-shaped, 4 mm long, exterior pubescent; filaments connate into a short tube, glabrous; ovaries 5-locular, glabrous. Capsules globose, 2.5 cm in diameter, 1-locular, 1-seeded.

Yunnan: Hekou, Dujiaotang, Liu Weixin 616.

Distribution: Yunnan.

27. *Camellia transarisanensis* (Hay.) Cohen-Stuart, Bull. Jard. Bot. Buitenz., ser. 3, 1: 320. 1919.

Thea parvifolia Hay., Icon. Pl. Formos. 3: 45. 1913, *non* Salisb.
Thea transarisanensis Hay., Icon. Pl. Formos. 5: 10. 1915.
Camellia parvifolia (Hay.) Cohen-Stuart, Meded. Proefst. Thee 40: 68. 1916.
Theopsis transarisanensis (Hay.) Nakai, Journ. Jap. Bot. 16: 707. 1940.

Branchlets pubescent. Leaves oblong-lanceolate, 4.5 cm long, 1.5 cm wide. Pedicels 3 mm long; calyx cup-shaped, 4 mm long, exterior pubescent; filament tube longer than the free filaments, glabrous; ovaries pubescent.

Taiwan: U. Mori 3549.

Distribution: Taiwan.

This species is close to *C. handelii*, but the calyx is shorter.

28. *Camellia fraterna* Hance, Ann. Sci. Nat. Paris 18: 219. 1862.

Thea rosaeflora var. *pilosa* Kochs in Engler, Bot. Jahrb. 27: 585. 1900.
Theopsis fraterna (Hance) Nakai, Journ. Jap. Bot. 16: 706. 1940.

Branchlets pubescent. Leaves elliptic, 4-8 cm long, 1.5-3.5 cm wide. Pedicels 3-4 mm long, sepals very velutinous, filament tube longer than free filaments, ovaries glabrous,

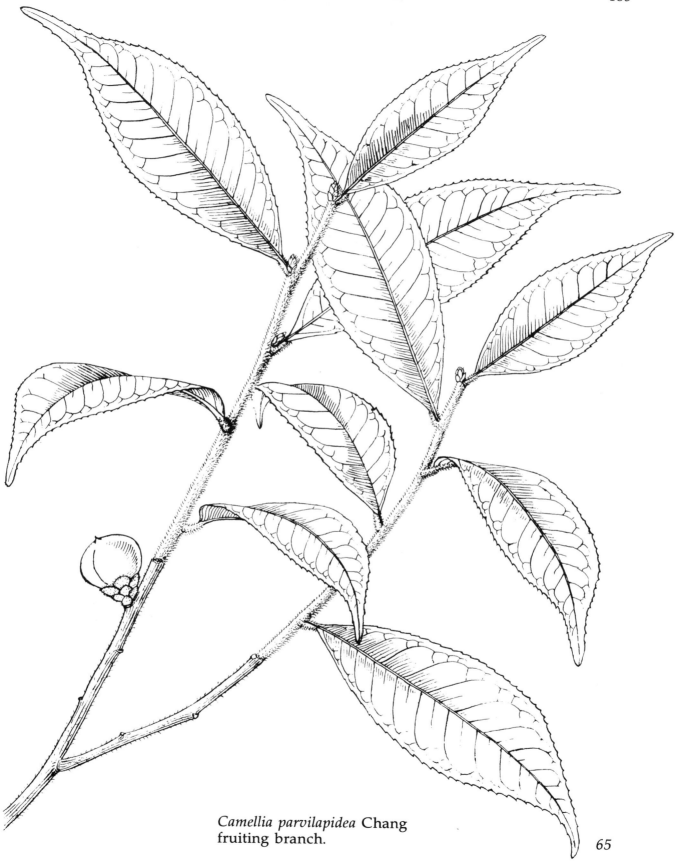

Camellia parvilapidea Chang fruiting branch.

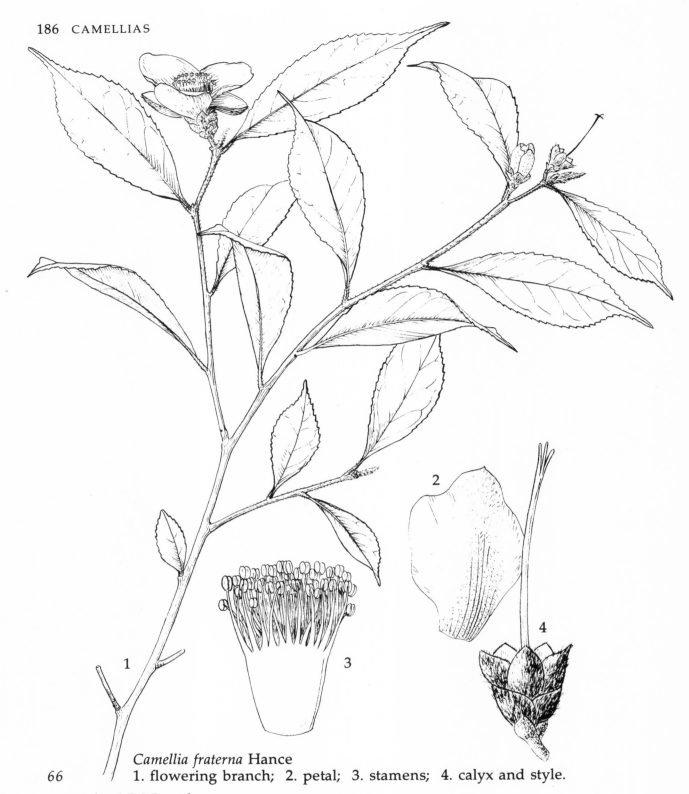

Camellia fraterna Hance
1. flowering branch; 2. petal; 3. stamens; 4. calyx and style.

styles 1.5-2.5 cm long.

Fujian: Tang Ruiyong 117, 433, 546, 547, 651, 1039.

Zhejiang: Longchuan, He Xianyu (H. Y. Ho) 3358; Zhuji, He Xianyu (H. Y. Ho) 2924; Tianmu Shan, Zhu Heqing 154; same loc., He Xianyu (H. Y. Ho) 21065; Changhua, He Xianyu (H. Y. Ho) 23789; Xitianmu, Deng Maobin 4520; Changhua, Deng Maobin 4541.

Camellia dubia Sealy
1. flowering branch; 2. sepals, petal and stamens; 3. calyx and style. *67*

Jiangsu: Xixing, Shen Jun 744; He Xianyu (H. Y. Ho) 3842.
Anhui: Huangshan, Wang Mingjin 3650; same loc., Fu Ligguo 584.
Distribution: Anhui, Jiangsu, Zhejiang, Fujian, Jiangxi.

29. *Camellia dubia* Sealy, Rev. Gen. Camellia, 68. 1958.
 This species is close to *C. fraterna* only the leaves are ovate, flowers purplish-red, and the pedicels and sepals glabrous or nearly glabrous.
 Sichuan: Yu Dejun (T. T. Yü) 304, 288.
 Hubei: Enshi, Li Hongjun 6084; Hefeng, Li Hongjun 6973, 8092.
 Jiangxi: Qiannan, Liu Xinqi 4390.
 Distribution: Sichuan, Hubei, Jiangxi.

30. *Camellia percuspidata* Chang, Acta Sci. Nat. Univ. Sunyatseni, monogr. ser. 1: 156. 1981.

A *C. cuspidata* foliis majoribus 8-12 cm longis 3-4.5 cm latis, pedicellis brevioribus, sepalis et petalis pubescentibus, filamentis connatis differt.

Arbor parva, ramulis glabris. Folia elliptica 8-12 cm longa 3-4.5 cm lata, apice caudata, caudo 1.5-2.3 cm longo, basi late cuneata, supra opaca subtus glabra, nervis lateralibus 5-7-jugis, margine serrulata, petiolis 5-7 mm longis. Flores terminales singuli vel bini, pedicellis 2 mm longis; bracteis 4 circ. 1-2 mm longis; sepalis 5 suborbiculatis 4-5 mm longis glabris; petalis 7 obovatis basi connatis pubescentibus, staminibus 2-seriatis exterioribus basi connatis breviter tubulatis glabris; ovariis glabris 3-locularibus, stylis apice 3-fidis. Capsula globosa 1.5 cm in diametro 1-locularis 3-valvata, valvis tenuibus, semina 1-2.

Small tree, branchlets glabrous. Leaves coriaceous, elliptic, 8-12 cm long, 3-4.5 cm wide, apices caudate with a 1.5-2.3 cm long cauda, bases broadly cuneate; shiny above, glabrous below; lateral veins 5-7 pairs, margins serrulate, petioles 5-7 mm long. Flowers 1-2 terminal, pedicels 2 mm long; bracts 4, 1-2 mm long; sepals 5, suborbicular, 4-5 mm long, pubescent; petals 7, obovoid, basally slightly connate, exterior pubescent; stamens in 2 series, outer whorl basally connate into a short tube, glabrous; ovaries glabrous, 3-locular; styles apices 3-cleft. Capsules globose, 1.5 cm wide, 1-locular, 3-valvate dehiscent, valves thin.

Yunnan: Tianxi, Sino-Soviet Expedition 2230 (holotype in KUN); Luchun, Huanglian Shan, elev. 1500-2000 m, Tao Deding 846, 1114, 1117.

31. *Camellia membranacea* Chang, Acta Sci. Nat. Univ. Sunyatseni, monogr. ser. 1: 159. 1981.

Species in sectione *Theopsi* ramulis juvenilibus glabris, foliis membranaceis ellipticis 7.5-10 cm longis, abrupte cuspidatis glabris, bracteis 5, sepalis suborbicularibus pubescentibus, capsulis 3-locularibus glabris distinguenda.

Frutex circ. 2.5 m altus, ramulis gracilibus glabris. Folia membranacea elliptica 7.5-10 cm longa 3-4.5 cm lata, supra opaca subtus glabra, apice abrupte attenuata caudata, caudo 1.5 cm longo, basi late cuneata, costa impressa puberula, nervis lateralibus 7-8-jugis utrinque visibilibus, retis nervorum utrinque inconspicuis, margine serrulata, petiolis 5-7 mm longis supra puberulis. Flores ignoti. Fructus soliatrii terminales, pedicellis 3-4 mm longis; bracteis 4-5 semiorbiculatis persistentibus puberulis; sepalis 5 suborbicularibus 4-5 mm longis pubescentibus; capsulis obovoideis 1.5 cm longis 3-locularibus glabris, semina solitaria in quoque loculo.

Shrub 2.5 m tall; branchlets slender, glabrous. Leaves membranous, elliptic, 7.5-10 cm long, 3-4.5 cm wide; apices abruptly attenuate, caudate with a 1.5 cm long cauda; not shiny above in the dry state; light-green below, glabrous; midveins impressed, with remnants of pubescence; lateral veins 7-8 pairs, slender, evident on both surfaces; reticulating veins obscure on both surfaces, margins densely serrate; petioles 5-7 mm long, rounded below, canaliculate above, puberulent. Flowers not seen. Capsules solitary, terminal, obovoid, 1.5 cm long, glabrous, 3-locular, 1 seed per locule, pedicels 3-4 mm long; bracts 4-5, semiorbicular, persistent, pubescent; sepals 5, suborbicular, 4-5 mm long, exterior pubescent.

Yunnan: Jinping, Dalaotang, elev. 1500 m, mountains in dense forests, shrub, 21 August 1951, Mao Pinyi (P. I. Mao) 446 (holotype in KUN).

32. *Camellia rosaeflora* Hook., Bot. Mag. 84: sub. pl. 5044. 1858.

Thea rosaeflora (Hook.) O. Kuntze, Rev. Gen. Pl., 64. 1891.

Branchlets pubescent. Leaves elliptic, 5-8 cm long, 2-2.5 cm wide, petioles 6-10 mm long. Pedicels 4-5 mm long, bracts 6-8; sepals 6-7 mm long, nearly free, pubescent; petals red, filament tube equal in length to the free filaments, ovaries glabrous.

Hubei: Yichang, October 1887, A. Henry 3374 (K).

Cult.: In hort. Kew, Feb.-March 1858 (holotype in K).

Distribution: Jiangsu, Zhejiang, Sichuan, Hubei.

This is a wild species. Sealy did not use Henry 3374 which was collected near Yichang in Hubei Province. This specimen was determined by Cohen-Stuart, and the author agrees with the determination.

33. *Camellia campanisepala* Chang, Acta Sci. Nat. Univ. Sunyatseni, monogr. ser. 1: 160. 1981.

A *C. fraterna* ramulis subglabris, sepalis calathiformibus longioribus et latioribus, filamentis subliberis, stylis liberis differt, a *C. cuspidata* sepalis petalisque pubescentibus, stylis 3 liberis recedit.

Frutex circ. 3 m altus, ramulis glabris vel primo puberulis glabrescentibus. Folia coriacea oblonga 6-9 cm longa 1.5-2.5 cm lata apice caudato-acuminata, caudo 1.5 cm longo, basi cuneata, supra in sicco olivaceoviridia nitida ad costam puberula subtus glabra, nervis lateralibus 8-11-jugis, supra visibilibus subtus inconspicuis, margine serrulata, petiolis 3-4 mm longis subtus glabris supra puberulis. Flores terminales albi 2.2 cm longi, pedicellis 5 mm longis; bracteis 5 ovatis vel ovoideis 1-3 mm longis pubescentibus; calyx campanulatus 8-9 mm longus basi connatus, limbis 5 triangulo-ovatis 4-5 mm longis pubescentibus; petalis 5, exterioribus l-2 late ovatis coriaceis basi leviter connatis pubescentibus, interioribus 3 obovatis 1.6 cm longis apice rotundatis pubescentibus; staminibus 1.5 cm longis basi 2-3 mm connatis, filamentis subliberis glabris; ovariis glabris, stylis 1.2 cm longis omnino liberis.

Shrub 3 m tall; branchlets glabrous, or initially slightly puberulent, soon glabrous. Leaves coriaceous, oblong, 6-9 cm long, 1.5-2.5 cm wide, apices caudate-acuminate with a 1.5 cm long cauda, bases cuneate; olive-green above in the dry state, shiny, glabrous; lateral veins 8-11 pairs, faint above, obscure below; margins serrulate, teeth separation 2-2.5 mm; petioles 3-4 mm long, upper region puberulent, lower region glabrous. Flowers terminal, 2.2 cm long; pedicels 5 mm long; bracts 5, ovate to ovoid, 1.3 mm long, pubescent; calyx campanulate, 8-9 mm long, pubescent; corolla white, 1.8 cm long; petals 5, outer 1-2 broadly ovoid, bases slightly connate, exterior pubescent, coriaceous; inner 3 petals obovoid, 1.6 cm long, apices rounded, exterior pubescent; stamens 1.5 mm long, basal 2-3 mm connate, free filaments glabrous; ovaries glabrous; styles 1.2 cm long, completely free, 3-parted, glabrous.

Guangdong: Renhua, Wanchi Shan, Shibixia Village, Zeng Huaide (W. T. Tsang) 26170 (holotype in SYS, isotype in A).

Sealy mentions this specimen but did not determine it because the flowers were not complete. The characterist of a 3-parted style is unprecedented in section *Theopsis*.

34. *Camellia lancilimba* Chang, Acta Sci. Nat. Univ. Sunyatseni, monogr. ser. 1: 161. 1981.

A *C. campanisepala* ramulis pubescentibus, foliis lanceolatis, pedicellis brevioribus 1.5-2.5 mm longis, bracteis paucioribus 2-3, sepalis brevioribus 5-6 mm longis, stylis connatis differt.

Frutex circ. 3 m altus, ramulis gracilibus pubescentibus. Folia tenuiter coriacea anguste lanceolata 4-6.5 cm longa 1-1.3 cm lata, apice caudato-acuminata, caudo 1-1.5 cm longo, basi late cuneata, supra nitidula ad costam puberula subtus glabra, nervis lateralibus 5-7-jugis utrinque inconspicuis, margine serrulata, petiolis 3-4 mm longis pubescentibus. Flores terminales ad axillares albi, pedicellis brevissimis; bracteis 3 late ovatis 1.5-2.5 mm longis puberulis; calyx breviter calathiformis 5 mm longus puberulis, limbis 5 ovatis apice obtusis; petalis 5-7 obovatis basi leviter connatis, exterioribus 2 coriaceis puberulis, ceteris obovatis 2 cm longis puberulis; staminibus petalis brevioribus exterioribus dimidis inferioribus connatis tubulatis glabris; ovariis glabris, stylis apice 3-fidis. Capsula globosa 1.6 cm in diametro 1-locularis, semina 1.

Shrub 3 m tall; branchlets slender, with grey pubescence. Leaves thinly coriaceous,

68 *Camellia campanisepala* Chang branch in bud.

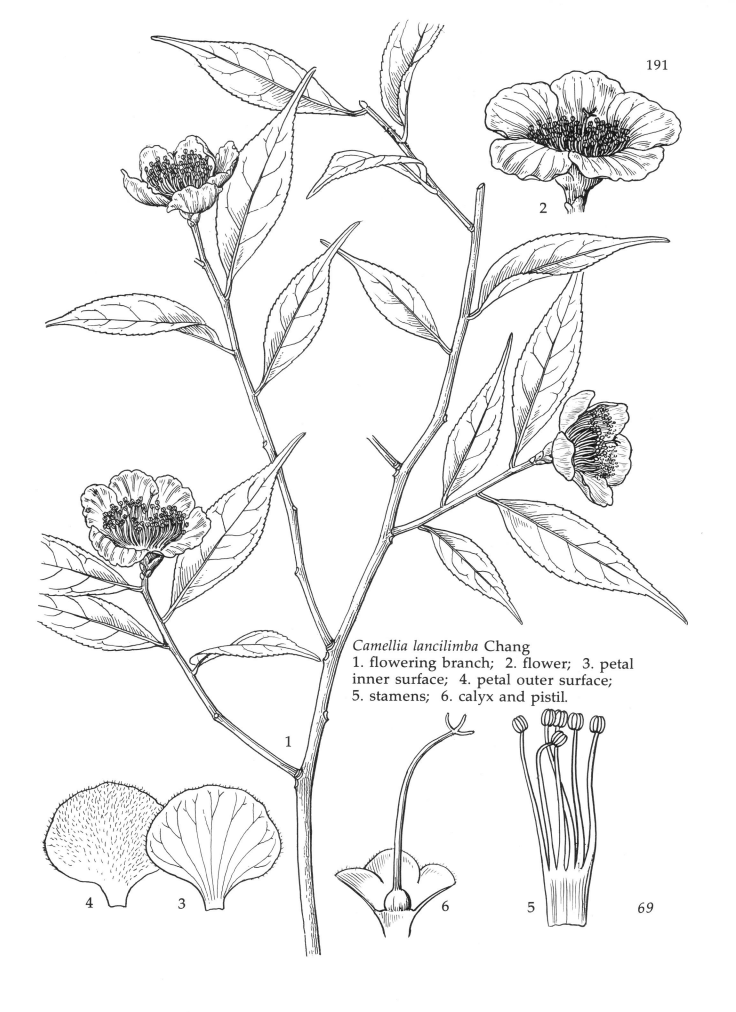

Camellia lancilimba Chang
1. flowering branch; 2. flower; 3. petal inner surface; 4. petal outer surface; 5. stamens; 6. calyx and pistil.

narrowly lanceolate, 4-6.5 cm long, 1-1.3 cm wide, apices caudate-acuminate, cauda 1-1.5 cm long, bases broadly cuneate or rounded; greenish-brown above, shiny, midvein puberulent; glabrous below; lateral veins 5-7 pairs, inconspicuous on both surfaces; margins densely serrate, teeth separation ca. 1 mm; petioles 3-4 mm long, pubescent. Flowers terminal to axillary, pedicels short; bracts 3, broadly ovate, 1.5-2.5 mm long, puberulent; calyx shallowly cup-shaped, 5 mm long, pubescent, upper half 5-parted; sepals ovate, apices obtuse; petals 5-7, broadly ovate, bases slightly connate, outer 2 coriaceous, puberulent, others obovate, 2 cm long, puberulent; stamens shorter than petals, outer filament whorl basally connate into a tube, upper half free, glabrous; ovaries glabrous, style apices shallowly 3-cleft. Capsules globose, 1.6 cm wide, 1-locular, 1-seeded.

Guangdong: Ruyuan, Liu Xinqi (S. K. Lau) 53950 (holotype in SCBI), 52887, 53583, 53128; Ruyuan, Wuzhi Shan, Huang Zhi (C. Wang) 44384, 43806, 44272; Qingluan, Huang Zhi (C. Wang) 30814; Ruyuan, Wuzhi Shan (Guangdong 737 1238); Liannan, Tan Peixiang 59574; Huaiji, Liu Yingguang 2592; Conghua, Zeng Huaide (W. T. Tsang) 25009.

Hunan: Mang Shan, Xiangguchang, Li Binggui 08; same loc., Liu Linhan (H. L. Li) 10234, 10235; Dao Xian, Tan Peixiang 61279.

Guangxi: Guangxi Agr. Inst. Forestry Dept. 780173.

SER. III.
Trichandrae Chang, Acta Sci. Nat. Univ. Sunyatseni, monogr. ser. 1: 163. 1981.

Filamenta basi connata tubifera pilosa, ovaria glabra.
Filaments connate into a tube, pilose; ovaries glabrous.
9 species distributed in the Chinese provinces south of the Changjiang (Yangtse River).
Type: *C. tsingpienensis* Hu

35. *Camellia tsingpienensis* Hu, Bull. Fan Mem. Inst. Biol. Bot. 8: 129. 1938.

35a. *Camellia tsingpienensis* Hu var. *tsingpienensis*
Branchlets pubescent. Leaves long-ovate, 5-8 cm long, 1.5-3 cm wide. Pedicels short; sepals glabrous, 2-3 mm long; filament tube short, free filaments pubescent; ovaries glabrous.
Yunnan: Jinping, Mao Pingyi (P. I. Mao) 3470, 2237; same loc., Cai, Xitao (H. T. Tsai) 52503, 52519 (lectotype in PE here designated, isolectotype in A).
Guangxi: Lingui, Feng Jinyong 1052.

35b. *Camellia tsingpienensis* var. *pubisepala* Chang, Acta Sci. Nat. Univ. Sunyatseni, monogr. ser. 1: 163. 1981.
A typo bracteis sepalis pubescentibus, petalis etiam pubescentibus, tubis filamentorum longioribus differt.
This variety differs from typical species by having pubescent bracts, sepals, and petals; filament tubes slightly longer.
Guangxi: Mubian, Dewo Village, Nantou Shan, small tree 4 m tall, flowers white, Gao Xipeng 55961 (holotype in SCBI); Napo Xian, Liang Shengye (S. Y. Liang) 721106.
Guizhou: Rongjian, Qian Nandui 2850; Guizhou Bot. Gard. 302; Guizhou For. Inst. 1221; Li Yongkang 78040.

36. *Camellia parviovata* Chang & Wang in Chang, Acta Sci. Nat. Univ. Sunyatseni, monogr. ser. 1: 163. 1981.
A *C. tsingpienensi* Hu foliis minoribus, pedicellis longioribus 5 mm longis, sepalis majoribus 5 mm longis, tubis filamentorum longioribus recedit.

Frutex vel arbor parva, circ. 2-3 m altus, ramulis hirtellis. Folia tenuiter coriacea ovato-lanceolata 4-6 cm longa, 1.5-2 cm lata, apice acuminata basi subrotundata, supra opaca subtus glabra, nervis lateralibus circ. 9-jugis, margine serrulata, petiolis 2-3 mm longis puberulis. Flores axillares vel terminales albi, pedicellis 4-5 mm longis; bracteis 4 dispersis glabris; sepalis basi connatis breviter calathiformibus 5 mm longis apice rotundatis; petalis 6-7 obovatis 1.5 cm longis utrinque puberulis; staminibus petala subaequalibus, inferioribus dimidis connatis tubulatis, filamentis liberis pubescentibus; ovariis glabris, stylis staminibuspaulo longioribus, apice 3-fidis.

Shrub or small tree 2-3 m tall, branchlets hirtellous. Leaves thinly coriaceous, ovate-lanceolate, 4-6 cm long, 1.5-2 cm wide, apices acuminate, bases subrounded; dull above, glabrous below; lateral veins ca. 9 pairs, margins serrulate; petioles 2-3 mm long, pubescent. Flowers white, pedicels 4-5 mm long; bracts 4, dispersed, glabrous; sepals 5 mm long, basally connate into a shallow cup, apices rounded, glabrous; petals 6-7, broadly ovate, 1.5 cm long, both surfaces somewhat puberulent; stamens ca. equal in length to petals, basally connate into a short tube, free filaments pubescent; ovaries glabrous; styles slightly exceeding stamen in length, shallowly 3-cleft.

Sichuan: Emei Shan, Daeshi, Yu Dejun (T. T. Yü) 304 (holotype in PE, isotype in A), 288.

37. *Camellia viridicalyx* Chang & Liang in Chang, Acta Sci. Nat. Univ. Sunyatseni, monogr. ser. 1: 164. 1981.

Species *C. tsingpienensem* affinis, sed differt foliis minoribus crassioribus, sepalis petalisque pubescentibus, tubis filamentorum longioribus.

Frutex circ. 1-2 m altus, ramulis brunneo-puberulis. Folia coriacea lanceolata vel ovato-lanceolata 3.5-5.5 cm longa 1-1.6 cm lata, apice caudato-acuminata, caudo 1-1.5 cm longo, basi late cuneate vel subrotundata, supra opaca ad costam puberula subtus glabra, nervis lateralibus circ. 7-jugis supra visibilibus subtus inconspicuis, margine serrulata, petiolis 2-3 mm longis puberulis. Flores terminales albi, pedicellis brevissimis; bracteis 4 ovatis 1.5-2.5 mm longis puberulis ciliatis; sepalis 5, circ. 5.5 mm longis basi leviter connatis, limbis ovoideis; petalis 5 obovatis basi connatis extus puberulis; staminibus petalis brevioribus, exterioribus inferioribus ½-⅔ connatis tubulatis, filamentis liberis pubescentibus; ovariis glabris, stylis stamina subaequantibus apice 3-fidis.

Shrub 1-2 m tall, branchlets brown-pubescent. Leaves coriaceous, lanceolate or ovate-lanceolate, 3.5-5.5 cm long, 1-1.6 cm wide, apices caudate-acuminate with a 1-1.5 cm long cauda, bases broadly cuneate or slightly rounded; green above in the dry state, not shiny, midveins puberulent; yellowish-brown below, glabrous; lateral veins ca. 7 pairs, visible above, obscure below; margins serrulate, teeth separation 1.5-2 mm; petioles 2-3 mm long, puberulent. Flowers terminal, pedicels short; bracts 4, ovate, 1.5-2.5 mm long, exterior puberulent, margins ciliate; sepals 5, 5.5 mm long, ovoid, bases slightly connate, exterior pubescent, margins ciliate; petals 5, obovate, bases slightly connate, 2 outer petals ovoid, exterior pubescent; stamens ca. equal to petal length, basal half of outer filament whorl connate, free filaments pubescent; ovaries glabrous; styles ca. equal in length to stamens, glabrous, apices shortly 3-cleft. Capsules globose, 1.8 cm in diameter, 3-valvate dehiscent, 2 seeded.

Guangxi: He Xian, Gupo Shan, by a mountain valley stream, Liang Shengye (S. Y. Liang) 5605229 (holotype in SYS), 5605227; He Xian, Liufu, Tiechangchong, Li Zhongdi 604118 (in fruit).

38. *Camellia lancicalyx* Chang, Acta Sci. Nat. Univ. Sunyatseni, monogr. ser. 1: 165. 1981.

Species *C. subglabrae*, assimilis, sed sepalis lanceolatis vel oblongis densius pubescentibus, petalis oblongis 3 mm latis differt.

Frutex, ramulis puberulis. Folia coriacea lanceolata 4-5 cm longa 1-1.3 cm lata, apice

70 1. *Camellia viridicalyx* Chang & Liang 2-3. *Camellia parvicaudata* Chang
1. flowering branch; 2. flowering branch; 3. calyx and pistil.

caudato-acuminata, basi obtusa vel subrotundata, supra nitida ad costam puberula subtus glabra, nervis lateralibus 7-8-jugis utrinque inconspicuis, margine crenulata, petiolis 2-3 mm longis pubescentibus. Flores singuli vel bini subterminaliter axillares albi, pedicellis brevissimis; bracteis 3 ovatis puberulis ad basi pedicelli dispositis; sepalis 5 lanceolatis vel oblongis 3 mm longis 1.5 mm latis subliberis apice subacutis vel obtusis puberulis; petalis 6 exterioribus 2 ovatis 6-8 mm longis puberulis, interioribus 4 oblongis 1-1.2 cm longis 2-3 mm longis basi 2 mm connatis extus serico-puberulis; staminibus 1 cm longis inferioribus dimidis connatis tubulatis glabris, superioribus liberis, filamentis pubescentibus; ovariis, stylis 1 cm longis apice 3-fidis.

Shrub, branchlets puberulent. Leaves coriaceous, lanceolate, 4-5 cm long, 1-1.3 cm wide, apices caudate-acuminate, bases cuneate or subrounded; dark-green above in the dry state, shiny, midveins puberulent or glabrous; light-green below, glabrous; lateral veins 7-8 pairs, obscure on both surfaces; margins serrulate, teeth separation 1-2 mm; petioles 2-3 mm long, pubescent. Flowers 1-2 in the terminal leaf axils, white, pedicels short; bracts 3, on the pedicel base, ovate, pubescent; sepals 5, lanceolate or oblong, 3 mm long, 1.5 cm wide, basally connate, apices subacute or obtuse, exterior pubescent; petals 6, outer 2 ovate, 6-8 mm long, 4-6 mm wide, exterior pubescent, inner 4 oblong, 1-1.2 cm long, 2-3 cm wide, bases connate for ca. 2 mm, exterior sericeous-puberulent; stamens 1 cm long, basal half of outer filament whorl connate into a short tube, glabrous, upper half of filament whorl free, pubescent; ovaries glabrous; styles 1 cm long, apices shallowly 3-cleft.

Guangxi: Wuming, Daming Shan, 17 October 1973, Liang Shengye (S. Y. Liang) 7301017 (holotype in SYS).

39. *Camellia parvicaudata* Chang, Acta Sci. Nat. Univ. Sunyatseni, monogr. ser. 1: 166. 1981.

Species floribus illis *C. nokoensem* assimilibus sed sepalis petalisque pubescentibus, corolla rubra, pedicellis brevioribus, bracteolis imbricatis distincta.

Frutex, ramulis puberulis. Folia coriacea oblonga vel oblongo-lanceolata 2.5-3.5 cm longa 8-13 mm lata, apice caudato-acuminata basi late cuneata, supra nitidula ad costam puberula, subtus glabra, nervis lateralibus 7-8-jugis, margine serrulata, petiolis 2 mm longis hirtellis. Flores axillares vel terminales, pedicellis 1-2 mm longis; bracteis 4-5 semilunatis vel late ovatis maximis 1.5 mm longis puberulis; sepalis 5 late ovatis 2.5 mm longis obtusis pubescentibus; corolla rubra 8-9 mm longa, petalis 5 basi connatis exterioribus 2 obovatis 6-7 mm longis coriaceis extus pubescentibus, interioribus 3 late obovatis 8-9 mm longis pubescentibus; staminibus 7 mm longis, exterioribus basi 2 mm connatis breviter tubulatis, filamentis liberis 4 mm longis pubescentibus; ovariis glabris, stylis 1.3 mm longis, apice 3-fidis. Capsula ignota.

Shrub, branchlets puberulent. Leaves coriaceous, minute, oblong or oblong-lanceolate, 2.5-3.5 cm long, 8-13 mm wide, apices caudate-acuminate, bases broadly cuneate or obtuse; olive-green above, shiny, midveins puberulent; yellowish-brown below, glabrous; lateral veins 7-8 pairs; margins densely serrulate, teeth separation 0.5-1 mm; petioles 2 mm long, hirtellous. Flowers terminal or axillary, pedicels 1-2 mm long; bracts 4-5, semiorbicular to broadly ovate, 0.5-1.5 mm long, imbricate, exterior pubescent; sepals 5, broadly ovate, 2.5 mm long, apices obtuse, exterior pubescent; corolla red, 8-9 mm long; petals 5, basally connate, adnate with the stamen filament tube, outer 2 petals obovoid, 6-7 mm long, exterior pubescent, coriaceous, inner 3 petals broadly ovate, 8-9 mm long, exterior pubescent, petal-shaped; stamens 7 mm long, basal 2 mm of outer whorl connate into a short tube, free filaments pubescent; ovaries glabrous; styles 1.3 cm long, glabrous, apices shallowly 3-cleft.

Guangxi: Wuming, Daming Shan, Liang Shengye (S. Y. Liang) 7307 (holotype in SYS); Dabsao, Huanglian Shan, elev. 1600 m, in dense forests, shrub, flowers red, Zhang Zhaoqian 13693; Lingui, flowers white, Zhong Jixin 808978.

40. *Camellia subglabra* Chang, Acta Sci. Nat. Univ. Sunyatseni, monogr. ser. 1: 167. 1981.

A *C. euryoidi* differt ramulis subglabris, floribus minoribus, filamentis pubescentibus, stylis pubescentibus apice profunde 3-fidis, lobis 3 mm longis; a *C. nokoensi* differt ramulis subglabris, tubis filamentorum longioribus, stylis pubescentibus apice profunde 3-fidis, lobis 3 mm longis.

Frutex circ. 2 m altus, ramulis puberulis mox glabrescentibus gracilibus. Folia coriacea lanceolata 3.5-4.5 cm longis 8-13 mm lata, apice acuminata vel caudata, caudo obtuso, basi late cuneata, supra nitida subtus muricatula, nervis lateralibus 5-6-jugis, margine serrulata, petiolis 1-2 mm longis puberulis. Flores terminales vel axillares 1.3 cm longi albi; pedicellis 3-4 mm longis glabris; bracteis 3-4 semilunatis miximis 1.5 mm longis glabris; sepalis 5 orbicularibus 2-2.5 mm longis glabris vel ciliatis; petalis 5 glabris exterioribus 1-2 ellipticis 7-8 mm longis basi leviter connatis, ceteris obovatis 1 cm longis, basi ad stamina adnatis; staminibus 8-9 mm longis exterioribus inferioribus ⅔ connatis tubulatis, tubis circ. 6 mm longis, filamentis liberis pilosis; ovariis glabris, stylis 7 mm inferioribus 4 mm connatis pilosis, superioribus 3-fidis lobis 3 mm longis. Capsula non visi.

Shrub 2 m tall; branchlets puberulent, soon glabrous. Leaves coriaceous, lanceolate, 3.5-4.5 cm long, 8-13 mm wide; apices acuminate or caudate, cauda obtuse; bases broadly cuneate or obtuse; shiny above in the dry state, mostly muricate below; lateral veins 5-6 pairs, margins densely serrulate; petioles 1-2 mm long, puberulent. Flowers terminal or axillary, 1.3 cm long; pedicels 3-4 mm long, glabrous; apices rounded; sepals 5, orbicular, 2-2.5 mm long, exterior glabrous, margins ciliate; petals 5, white, glabrous, outer 1-2 elliptic, 7-8 mm long, barely connate, remaining 3 obovate, 1 cm long, basal ca. 2 mm adnate with stamens; stamens 8-9 mm long, outer filament whorl connate into a short tube, tube ca. 6 mm long, free filaments pilose; ovaries glabrous; styles 7 mm long, basal 4 mm connate, pilose, upper 3 mm 3-cleft, glabrous.

Guangdong: Dinghu Shan, Jilong Shan, Li Qijing 106 (holotype in SYS).

41. *Camellia nokoensis* Hay., Icon. Pl. Formos. 8: 10. 1919.
Thea nokoensis (Hay.) Mak. & Nem., Fl. Jap., ed. 2, 745. 1931.
Theopsis nokoensis (Hay.) Nakai, Journ. Jap. Bot. 16: 706. 1940.
Camellia euryoides var. *nokoensis* Liu, Lign. Pl. Taiwan, 421. 1976.
Taiwan: Kaneshira et Sasaki 63.
Jiangsu: Y. C. Yang 468.
Distribution: Jiangsu and Taiwan.

C. nokoensis has pubescent filaments, pedicels relatively short, in these characters it differs from *C. euryoides*.

42. *Camellia tsofuii* Chien, Contrib. Biol. Lab. Sci. Soc. China Bot. 12: 91. 1939.

Branchlets pubescent. Leaves thinly coriaceous, ovate-oblong, 3-4.5 cm long. Pedicels slender, 5 mm long; bracts 4, minute; sepals 1.5 mm long, ciliate; petals 5, 7-15 mm long; sepals 1.5 cm long; filaments pubescent, lower ⅓ connate; ovaries glabrous. Capsules globose, 1-1.3 cm in diameter.

Sichuan: Emei Shan, Yang Guanghui 57038, 57152; Nanchuan, Li Guofeng 60546.
Hunan: Acad. Sin. Hunan Investigation Team, July 1953, 309.
Jiangxi: Ping Village, Acad. Sin. Jiangxi Team 654, 2575, 2635.
Distribution: Sichuan, Hunan, Jiangxi.

71 1-3. *Camellia trigonocarpa* Chang 4-5. *Camellia villicarpa* Chien 6-7. *Camellia tsofuii* Chien
1. flowering branch; 2. stamens; 3. calyx and style; 4. fruiting branch; 5. calyx; 6. flowering branch; 7. calyx and style.

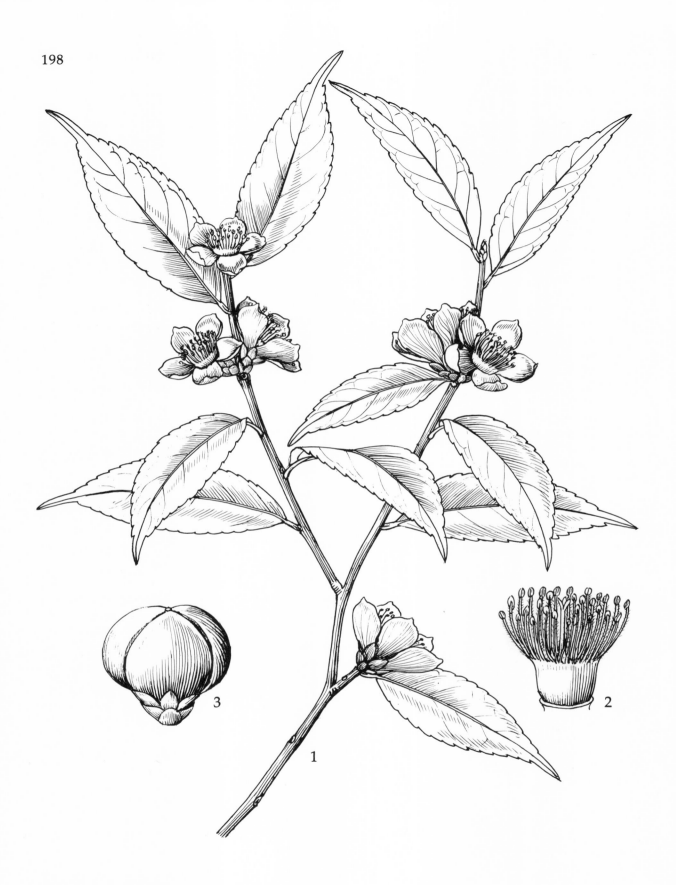

72 *Camellia trichandra* Chang
1. flowering branch; 2. stamens and style; 3. capsule.

43. *Camellia trichandra* Chang, Acta Sci. Nat. Univ. Sunyatseni, monogr. ser. 1: 168. 1981.

A *C. tsofuii* ramulis subglabris, foliis longioribus; floribus subsessilibus, petalis oblongis, staminibus brevioribus, tubis filamentorum longioribus, stylis longioribus, capsulis compresse tricoccis differt.

Frutex, ramulis puberulis vel glabrescentibus. Folia tenuiter cariacea obovato-oblonga 4-7.5 cm longa 1.5-2.5 cm lata, apice caudato-acuminata, basi cuneata, utrinque glabra, nervis lateralibus 6-8-jugis, margine serrulata, petiolis 3-4 mm longis puberulis. Flores terminales albi, pedicellis 1-2 mm longis; bracteis 4 ovatis 1 mm longis; sepalis 5 ovatis 2 mm longis apice obtusis ciliatis; petalis 5 oblongis 1.3-1.8 cm longis 5-7 mm latis basi leviter connatis glabris; staminibus 1.5 cm longis, inferioribus connatis, tubo filamentorum 8-9 mm longo, superioribus liberis pubescentibus; ovariis glabris, stylis 1.4 cm longis. Capsula compresse triangulo-globosa 3-locularis 1 cm lata, semina singula in quoque loculo.

Shrub, branchlets glabrescent or puberulent. Leaves thinly coriaceous, obovate-oblong, 4-7.5 cm long, 1.5-2.5 cm wide, apices caudate-acuminate, bases cuneate, lateral veins 6-8 pairs, margins serrate; petioles 3-4 mm long, pubescent. Flowers sub-terminal, pedicels 1-2 mm long, bracts 4; sepals ovate, 2 mm long, apices obtuse, ciliate; petals 5, oblong, 1.3-1.8 cm long, 5-7 mm wide, basally connate, glabrous; stamens 1.5 cm long, filament tube 8-9 mm long, free filaments pubescent; ovaries glabrous, styles 1.4 cm long. Capsules oblate-tricoccus, 3-locular, 1 cm wide, 1 seed per locule.

Guangxi: Lingle, Lao Shan, 19 December 1957, Zhang Zhaoqian 11125 (holotype in SCBI); same loc., Fucheng, Qinglong Shan, Zhang Zhaoqian 10486; Baise, Siqu, Dawan Shan, 27 December 1955, Acad. Sin. Guangxi Investigation Team, Baise Team 1816.

Yunnan: Funing, Wang Qingwu 88435.

SECTION XX.
Eriandria Cohen-Stuart, Meded. Proefst. Thee 40: 69. 1916.

Camellia sect. *Camelliopsis* (Pierre) Sealy, Rev. Gen. Camellia, 97. 1958, *nom illegit*.
Thea sect. *Camelliopsis* Pierre, Fl. For. Cochinchine 2: sub. pl. 119. 1887.

Flowers small or medium, 1-2 axillary, white, rarely red, short pedicellate; bracts 2-5, minute, persistent; sepals 5-6, basally connate; stamens in 1-2 series, often connate into a filament tube, often pubescent, anthers basifixed; ovaries pubescent, 3-locular; styles connate, apices 3-cleft. Capsules 1-locular, without a columella, 1 seeded.

14 species; mostly in China, some species occur in Indo-China and India.

Type: *C. caudata* Wall.

KEY TO SECT. *ERIANDRIA*

1. Filaments glabrous, ovaries pubescent.
    2. Filaments free, rarely bases adnate with petals; sepals pubescent.
        3. Sepals 2 mm long, leaves 2-4 cm long, flowers white . . . . . 1. *C. villicarpa* Chien
        3. Sepals 5-6 mm long, leaves 7 cm long, flowers purplish-red . . . . 2. *C. cratera* Chang
    2. Basal half of filaments connate into a short tube, sepals glabrous or pubescent.
        4. Sepals nearly free, exterior glabrous or ciliate; fruit globose, leaves elliptic, inside of seed coat glabrous . . . . 5. *C. trigonocarpa* Chang
            5. Sepals not ciliate, bracts completely imbricate on pedicels, petals pubescent, leaf serrations obtuse . . . . 3. *C. punctata* (Kochs) Cohen-Stuart
            5. Sepals long-ciliate, bracts dispersed, sepals glabrous, leaf serrations sharp. . . . . 4. *C. lawaii* Sealy

>   4. Sepals connate or cup-shaped, exterior pubescent; fruit triangular-ellipsoid, inside of seed coats pubescent, leaves narrowly oblong..5. *C. trigonocarpa* Chang
> 1. Filaments and ovaries pubescent.
>   6. Sepals orbicular or ovate, 2-5 mm long.
>   7. Leaves serrate, lateral veins 8-9.
>   8. Leaves oblong-lanceolate, bases rounded, often pubescent below.
>   9. Leaves 4-12 cm long, 2-3 cm wide.
>   10. Pedicels 3 mm long, leaves 6-10 cm long ....6. *C. cordifolia* (Metc.) Nakai
>   10. Pedicels shorter, leaves 4-6 cm long.......7. *C. wenshanensis* Hu
>   9. Leaves 3-5 cm long, 1-1.3 cm wide ....8. *C. melliana* Hand.-Mazz.
>   8. Leaves oblong or oblong-lanceolate, bases cuneate, glabrous below.
>   11. Pedicels 6-9 mm long; leaves submembranous, longer than 10 cm ....9.*C. candida* Chang
>   11. Pedicels 6-9 mm long; leaves coriaceous, shorter than 10 cm.
>   12. Flowers 1-2 cm long; fruit 1-1.3 cm wide, 1-locular, pericarp shell-thin.
>   13. Sepals 1.5-2.5 mm long, free; filaments villous, pedicels 3-4 mm long.......................10. *C. caudata* Wall.
>   13. Sepals 5-6 mm long, bases connate or cup-shaped; filaments rarely pilose, pedicels 1-2 mm long .... 11. *C. assimiloides* Sealy
>   12. Flowers 3 cm long; fruit 1.5-2 cm wide, 2-3-locular, pericarp 1.5 mm thick; sepals 4-5 mm long, fruit pedicels 5-7 mm long ....12. *C. assimilis* Champ. ex Benth.
>   7. Leaves entire, lateral veins 9-10 pairs...............13. *C. edentata* Chang
>   6. Sepals filiform-lanceolate, 1-1.5 cm long, abundantly pilose; leaves lanceolate, bases rounded.........................14. *C. salicifolia* Champ. ex Benth.

1. **Camellia villicarpa** Chien, Contrib. Biol. Lab. Sci. Soc. China Bot. 12: 99. 1939.
   Camellia obscurinervis Tsai & Feng, Acta Phytotax. 1: 190. 1951.

   Branchlets pubescent. Leaves elliptic to oblong, 2-5 cm long, 1-1.5 cm wide. Pedicels 3-4 mm long, bracts 3-4; sepals only ciliate, 2 mm long; stamens free, glabrous; ovaries pubescent, styles free.

   Sichuan: Emei Shan, Li Guofeng 53341; same loc., Xiong Jihua, Zhang and Jiang *et al.* 32881, 33308, 33319; Nanchuan, Li Guofeng 64923; Qingcheng Shan, Li Xin 47174; Emei Shan, Yang Guanghui 56999; same loc., He Wenrong 839 (holotype of *C. obscurinervis* in KUN).

   The leaf blades of Xiong, Zhang and Jiang *et al.* 32881 are specially small, less than 2 cm long.

2. **Camellia cratera** Chang, Acta Sci. Nat. Univ. Sunyatseni, monogr. ser. 1: 170. 1981.

   Species *C. caudatae* affinis, a qua differt sepala majoribus connatis cyathiformibus, petalis rubris, filamentis liberis, stylis glabris; a *C. campanisepala* et *C. lancilimba* praesertim filamentis liberis, petalis rubris, ovario pubescenti distincta.

   Frutex vel arbor parva circ. 5 m altus, ramulis griseo-pubescentibus. Folia coriacea anguste oblonga vel oblongo-lanceolata 5-8.5 cm longa, 1.5-2.5 cm lata, apice acuminato-caudata basi cuneata, supra opaca ad costam puberula subtus glabra, nervis lateralibus 6-7-jugis inconspicuis, margine serrulata, petiolis 2-4 mm longis. Flores axillares vel terminales, pedicellis 5-6 mm longis; bracteis 5 maximis 2-2.5 mm longis; calyx calathis 5-6 mm longis pubescentibus, sepalis 5 semilunatis 2-2.5 mm longis; corolla punicea, petalis 6-7 exterioribus obovatis 1 cm longis interioribus obovatis 1.5 cm longis pubescentibus; staminibus 1.2-1.4 cm longis liberis, filamentis puberulis; ovariis pilosis, stylis glabris 1.6 cm longis apice 3-fidis lobis 2-2.5 mm longis. Capsula ignota.

Shrub or small tree 5 m tall; branchlets greyish-brown pubescent. Leaves coriaceous, narrowly oblong or oblong-lanceolate, 5-8.5 cm long, 1.5-2.5 cm wide, apices caudate-acuminate with a 1.5-2 cm long cauda, bases narrowly cuneate; greyish-green above in the dry state, slightly dull, midveins puberulent; pubescent below; lateral veins 6-7 pairs, extremely obscure; margins serrulate, teeth separation 2-3 mm; petioles 2-4 mm long, pubescent. Flowers axillary or terminal; pedicels 5-6 mm long, pubescent, upper region larger; bracts 5, lower three minute, semiorbicular, dispersed along the pedicel, upper 2 tightly appressed to the calyx tube, semiorbicular, 2-2.5 mm long, more or less pubescent; calyx connate into a cup, 5-6 mm long, pubescent, upper region divided into 5 sepals, semilunate, 2-2.5 mm long, apices rounded; corolla purplish-red; petals 6-7, outer 2 obovoid, 1 cm long, pubescent, inner 4-5 obovate, 1.5 cm long, exterior pubescent, bases adnate with stamens for ca. 3 mm; stamens 1.2-1.4 cm long, exterior basally adnate with petals, remainder free, filaments puberulent; ovaries pubescent; styles pubescent, 1.6 cm long, apices shallowly 3-cleft, clefts 2-2.5 mm long.

Guangdong: Yangshan Xian, Wuyuan Village, Nanmutang, mountainous valley, tree 5 m tall, flowers purplish-red, Deng Liang 1067 (holotype in SCBI); Zijin to Huidong, Wuqinzhang, Wei Zhaofen (C. F. Wai) 120992.

Jiangxi: Ruijin, Hongmendi to Fujian Changting, Lushan Bot. Gard. no. 3629.

3. *Camellia punctata* (Kochs) Cohen-Stuart, Meded. Proefst. Thee 40: 68. 1916.
Thea punctata Kochs in Engler, Bot. Jahrb. 27: 584. 1900.
Branchlets pubescent. Leaves elliptic, thinly coriaceous, 4-7 cm long, 1.5-2 cm wide. Pedicels 2 mm long, bracts 5; sepals 5, 5 mm long, glabrous; stamens glabrous, filament tube longer than half the length; ovaries pubescent, styles glabrous.

Sichuan: Le Shan, Fang Wenpei (W. P. Fang) 2284; same loc., Zhou Hechang 12911; Jie Lunying 748; Zhao Rukai 29.

Distribution: Sichuan.

4. *Camellia lawii* Sealy, Kew Bull. 1951(2): 180. 1951.
Branchlets pubescent. Leaves elliptic 4-8 cm long, 2-2.7 cm wide. Pedicels 1-2 mm long, bracts 4; sepals 3.5 mm long, ciliate; stamens with a short filament tube, ovaries pubescent.

Sichuan: Emei Shan, Zhou Hongfu 26932; Huarong Shan, Yang Xianjin 4184; Beipei, Jinyun Shan, Y. W. Lau 335 (holotype in K, isotype in SCBI).

Distribution: Sichuan, Jiangsu.

5. *Camellia trigonocarpa* Chang, Acta Sci. Nat. Univ. Sunyatseni, monogr. ser. 1: 173. 1981.
A *C. caudata* differt ramulis subglabris, filamentis glabris, capsulis triangulo-ellipsoideis; a *C. cratera* ramulis subglabris, filamentis inferioribus connatis glabris, capsulis triangulo-ellipsoideis recedit.

Arbor parva circ. 4 m alta, ramulis primo puberulis mox glabrescentibus. Folia coriacea anguste oblongo-lanceolata 5.5-8 cm longa 1.5-2 cm lata, apice caudata, basi cuneata, supra nitida, basi glabra, nervis lateralibus 6-jugis inconspicuis, margine serrulata, petiolis 2-4 mm longis pubescentibus. Flores terminales vel axillares albi, pedicellis 1-2 mm longis; bracteis 1.5-2 mm longis; calyx calthus 5 mm longus pubescens, sepalis 5 orbiculatis; petalis 5 basi leviter connatis extus pubescentibus; staminibus petalis brevioribus, filamentis basi connatis tubulatis, superioribus liberis glabris; ovariis pilosis, 1.4 cm longis pubescentibus apice 3-fidis. Capsula triangulo-ellipsoidea 2 cm longa 1.3 cm lata pubescens, 1-locularis, 3-valvata dehiscent, valvis 1-1.5 mm crassis, semina singula in quoque loculo, pedicellis 2-3 mm longis.

Small tree 4 m tall; branchlets puberulent, soon glabrous, older branches dull brown. Leaves coriaceous, narrowly oblong-lanceolate, 5.5-8 cm long, 1.5-2 cm wide, apices caudate, bases cuneate, olive-green above in the dry state, shiny, glabrous, muricate;

lateral veins ca. 6 pairs, margins serrulate; petioles 2-4 mm long, pubescent. Flowers axillary or terminal, pedicels 1-2 mm long; bracts 5, 2 at the pedicel base, 2 in the pedicel middle, semilunar, 1.5 mm long, pubescent, remaining 1 bract near the calyx, 2 mm long; calyx cup-shaped, 5 mm long, pubescent, upper region 5-cleft, clefts orbicular, apices rounded; petals 5, bases slightly connate, outer margins pubescent; stamens shorter than petals; filaments basally connate into a short tube, free filaments glabrous; ovaries pubescent; styles 1.4 cm long, pubescent, apices 3-cleft. Capsules triangular-oblong, 2 cm long, 1.3 cm wide, pubescent, 1-locular, 2-valvate dehiscent, valves 1-1.5 mm thick, woody; seeds 1, 1.5 cm long, 1 cm wide, seed coat black; persistent calyx cup-shaped, pubescent; fruit pedicel 2-3 mm long.

Guangdong: Lechang, Datiaokeng, Chen Nianqu (N. K. Chun) 42388.

Hunan: bordering Guangdong Yuyuan, Mang Shan, Pinghang, Sanping, elev. 850 m, evergreen forest, Huang Maoxian 112912 (holotype in SCBI).

This species differs from *C. caudata* and *C. cratera* by having a long cylindrical fruit with three ridges. In addition with *C. caudata* the filaments are pubescent, and with *C. cratera* the outer filament whorl is connate.

6. ***Camellia cordifolia*** (Metc.) Nakai, Journ. Jap. Bot. 16: 692. 1940.

Thea cordifolia Metc., Lingnan Sci. Journ. 11: 17. 1932.

Branchlets pubescent. Leaves ovoid, bases rounded, somewhat pubescent. Flowers short pedicelate, bracts 4-5; sepals 5, ovoid, pubescent; petals pubescent, apices 3-cleft. Capsules 3-locular, sometimes 1-2-locular.

Guangdong: Beijiang, Chen Huanyong (W. Y. Chun) 6098 (holotype in SYS, isotype in A); Lian Xian, Gao Ximing 50900; Wengyuan, Liu Xinqi (S. K. Lau) 806; Beijiang, Huang Zhi (C. Wang) 31429, 31482; Lechang, Zeng Huaide (W. T. Tsang) 2087.

Guangxi: He Xian, Liang Shengye (S. Y. Liang) 6505230.

Hunan: Yongxing Xian, Liyutang Commune, Huaping Brigade, Yaojiaping, Lei Yabi s.n.

Jiangxi: Qiannan, Liu Xinqi (S. K. Lau) 4091.

Taiwan: Hayasi s.n. (former Taiwan For. Inst. no. 16845).

Distribution: Guangdong, Guangxi, Jiangxi, Taiwan, Hunan.

The Taiwan specimen has membranous leaves. Among the Guangdong specimens some also have membranous leaves and are very similar to *C. salicifolia*. These plants differ from the typical *C. cordifolia* which has leathery leaves, and this may indicate some hybridization between these two species.

7. ***Camellia wenshanensis*** Hu, Bull. Fan Mem. Inst. Biol. Bot. 8: 130. 1938.

Guizhou: Anlong, Shipan Commune, Zhang Zhisong and Zhang Yongtian 3007.

Yunnan: Wen Shan, Cai Xitao (H. T. Tsai) 51612 (holotype in PE, isotype in A and K).

This species is close to *C. cordifolia* only the leaves are smaller, 4-6 cm long, flowers sessile; filament tube short, less than ⅔ the stamen length.

8. ***Camellia melliana*** Hand.-Mazz., Anz. Akad. Wiss. Math. Nat. Wien 59: 58. 1922.

Thea melliana (Hand.-Mazz.) Merr., Lingnan Sci. Journ. 7: 315. 1931.

Guangdong: Zengcheng, Zeng Huaide (W. T. Tsang) 20079; Longmen, Feng Qin (H. Fung) 18702; same loc., Li Xuigen 200269; Conghua, Zeng Pei 137.

Distribution: Guangdong.

This species is very close to *C. cordifolia* only the leaf blades are narrowly oblong, 3-5 cm long, 1-1.3 cm wide.

9. ***Camellia candida*** Chang, Acta Sci. Nat. Univ. Sunyatseni, monogr. ser. 1: 177. 1981.

A *C. caudata* foliis submembranaceis, pedicellis longioribus 6-9 mm longis, bracteis pluribus 5-6, antheris subulatis differt.

Arbor parva 3-8 m alta, ramulis pubescentibus. Folia membranacea oblongo-

1-3. *Camellia cordifolia* (Metc.) Nakai 4-6. *Camellia caudata* Wall.
1. flowering branch; 2. flower; 3. calyx and style; 4. flowering branch;
5. petals and stamens; 6. calyx and style.

lanceolata 8-12 cm longa 2.5-3.5 cm lata, apice caudato-acuminata basi cuneata, supra in sicco opaca, subtus plus minusve puberula, nervis lateralibus 7-8-jugis, margine serrulata, petiolis 4-6 mm longis puberulis. Flores axillares vel terminales albi 2.5 cm in diametro; pedicellis 6-9 mm longis puberulis; bracteis 5-6 dispersis; sepalis 5 ovoideis 2.5-3 mm longis puberulis basi connatis apice obtusis; petalis 5-6 obovatis 1.3 cm longis extur puberulis; staminibus petalis paulo longioribus, filamentis inferioribus dimidis connatis tubulatis, superioribus liberis pilosis; ovariis pilosis, stylis 1.2 cm longis pubescentibus apice 3-fidis. Capsula globosa 2 cm in diametro pubescens, semina singula in quoque loculo.

Small tree, branchlets slender with pubescence. Leaves submembranous, oblong-lanceolate, 8-12 cm long, 2.5-3.5 cm wide, apices caudate, bases cuneate, shiny above, somewhat pubescent below, lateral veins 7-8 pairs, margins serrulate; petioles 4-6 mm long, pubescent. Flowers white; pedicels 6-9 mm long, pubescent; bracts 5-6, minute; sepals 5, ovoid, 2.5-3 mm long, puberulent, apices obtuse, basally connate; petals 5-6, obovoid, 1.3 cm long, puberulent; stamens slightly exceeding petals in length, basal half connate into a short tube, free filaments pilose, anthers apically acute; ovaries pubescent; styles 1.2 cm long, pubescent, shallowly 3-cleft. Capsules globose, 2 cm in diameter, 1-locular, pubescent, seeds 1-2.

Yunnan, Malipo, Huangjinyin, Wang Qiwu (C. W. Wang) 83855 (holotype in KUN); same loc., Feng Guomei (K. M. Feng) 12995, 13488.

10. *Camellia caudata* Wall., Pl. Asiat. Rar. 3: 36. 1832.

Camellia axillaris Griff., Journ. Travels, 38, 45. 1847, *auct. non* Roxb. ex Ker, *nec* Roxb. ex Wall.

Thea caudata (Wall.) Seem., Trans. Linn. Soc. London 22: 348. 1859.
Camellia gracilis Hemsl., Ann. Bot. 9: 146. 1895.
Camellia buisanensis Sasaki, Trans. Nat. Hist. Soc. Formosa 21: 222. 1931.
Thea buisanensis (Sasaki) Metc., Lingnan Sci. Journ. 12: 180. 1933.
Camelliastrum caudatum (Wall.) Nakai, Journ. Jap. Bot. 16: 701. 1940.
Camelliastrum gracile (Hemsl.) Nakai, Journ. Jap. Bot. 16: 701. 1940.
Camelliastrum buisanense (Sasaki) Nakai, Journ. Jap. Bot. 16: 700. 1940.
Camellia caudata var. *gracilis* (Hemsl.) Yamamoto ex Keng, Taiwania 1: 234. 1950.
Theopsis caudata (Wall.) Hu, Acta Phytotax. Sin. 8: 166. 1963.

Branchlets glabrous. Leaves narrowly oblong to broadly-elliptic, 5-13 cm long, 1-4 cm wide, caudate-acuminate. Pedicels 3 mm long, bracts 3-5; sepals 2 mm long, pubescent; petals pubescent; filament tube long, free filaments pubescent; ovaries and styles pubescent.

Yunnan: Jinping, Mao Pingyi (P. I. Mao) 3208.

Guangxi: Yao Shan, Xin Shuzhi 3535; Shiwan Dashan, Zeng Huaide (W. T. Tsang) 2276, 22602, 23878, 24163, 24813; Zhaoping, Liang Shengye (S. Y. Liang) 6505170.

Guangdong: Luofu Shan, F. P. Metcalf 17465, 17640, 17837, E. D. Merrill 10228; Wenyuan, Liu Xinqi (S. K. Lau) 2632; Mei Xian, Zeng Huaide (W. T. Tsang) 21436; Huaiji, Zeng Huaide (W. T. Tsang) 22835; Fuyuan, Huang Zhi (C. Wang) 42701, 30067; Xinyi, Huang Zhi (C. Wang) 30948, 37753.

Hong Kong: Daao Island, Jiang Ying (Y. Tsiang) 3245.
Hainan: Bawangling, Zeng Pei 13461; Baoxiang, Hou Kuanzhao 73863, 73697.
Zhejiang: Zhihai, Zhong Buqin (P. C. Tsoong) 1050.
Vietnam: Dahuangmao Shan, Zeng Huaide (W. T. Tsang) 27158, 27314.
Distribution: Yunnan, Guangdong, Guangxi, Hainan, Taiwan, Zhejiang, Vietnam, Burma, Bhutan, India.

11. *Camellia assimiloides* Sealy, Kew Bull. 1949(2): 215. 1949.

Branchlets pubescent. Leaves oblong, caudate. Flower shape similar to *C. caudata* only the sepals are larger, 5-6 mm long, lower half connate into a cup; pubescence of filaments very sparsely pilose.

Guangdong: Beijiang, Banlingzi, Chen Huanyong (C. Y. Chen) 5906 (holotype in E); Lechang, Chen Niangou 42388.

12. *Camellia assimilis* Champ. ex Benth. in Hook., Journ. Bot. 3: 309. 1851.
Thea assimilis (Champ.) Seem., Trans. Linn. Soc. 22: 349. 1859.
Camelliastrum assimile (Champ.) Nakai, Journ. Jap. Bot. 16: 700. 1940.
Guangdong: Lechang, Guikeng, Gao Xipeng 51108; Haifeng, Gaotan, Wuzizhang, SCBI Plant Geogr. Dept. 7602.
Hong Kong: Champion 65 (photo of type); Daao Island, Jiang Ying (Y. Tsiang) 3214.
Guangxi: Rong Xian, no collector 41 (GXFI no. 7500127).
Distribution: Islands of the coast of central Guangdong and Guangxi.

The leaf shape of this is very similar to *C. caudata*, but the flowers are 3 cm long, filament tube ¾ the filament length, pedicels shorter, sepals longer, capsules 2-3 locular.

13. *Camellia edentata* Chang, Acta Sci. Nat. Univ. Sunyatseni, monogr. ser. 1: 178. 1981.

Species *C. caudata* affinis, a qua differt foliis integris nervis lateralibus pluribus, bracteis sepalisque majoribus acutis, tubis filamentorum brevioribus, filamentis sparse pubescentibus.

Frutex, ramulis pubescentibus. Folia coriacea oblonga 5-7.5 cm longa 1.5-2.3 cm lata, apice acuminata, basi cuneata, supra opaca ad costam puberula subtus glabra, nervis lateralibus 10-12-jugis supra plus minusve visibilibus subtus inconspicuis, integra, petiolis 3-5 mm longis pubescentibus. Flores terminales vel axillares, pedicellis brevissimis; bracteis 5 semilunatis vel orbiculatis 1-3 mm longis pubescentibus; sepalis 5 ovatis subliberis 4-5 mm longis pubescentibus, apice acutis; petalis albis 7, 1-1.2 cm longis, exterioribus 2 suborbiculatis coriaceis 8 mm longis, interioribus 5 obovatis basi 2 mm connatis subcoriaceis, pubescentibus; staminibus 1 cm longis inferioribus dimidis connatis, tubo filamentorum 5 mm longo, superioribus liberis 5 mm longis sparse puberulis; ovariis pilosis, stylis 8-10 mm longis pubescentibus apice 3-fidis. Capsula globosa 1.5 cm in diametro, pericarpio tenui, 1-locularis; semina singula globosa 1.2 cm diam.

Shrub, branchlets grayish-puberulent. Leaves coriaceous, oblong, 5-7.5 cm long, 1.5-2.3 cm wide, apices acuminate, bases cuneate; green above, greyish-green in the dry state, not shiny, midvein puberulent; same color below, glabrous; lateral veins 10-12 pairs, somewhat visible above, obscure below; entire; petioles 3-5 mm long, pubescent. Flowers terminal or axillary, pedicels very short; bracts 5, semilunar to orbicular, 1-3 mm long, pubescent; sepals 5, ovate, completely free, 4-5 mm long, pubescent, apices acute; petals white, 7, 1-1.2 cm long, outer 2 petals suborbicular, coriaceous, 8 mm long, pubescent, inner 5 petals ovoid, basally connate for ca. 2 mm, thicker towards the center, coriaceous, margins thin, pubescent; stamens 1 cm long, filament tube ca. 5 mm long, free filaments 5 mm long, sparsely pubescent; ovaries pubescent; styles pubescent, apices shallowly 3-cleft. Capsules globose, 1.5 cm in diameter, 1-locular, 1 seeded, pericarp thin; seeds globose, 1.2 cm in diameter, seed coat black.

Guangdong: Enping, Qingwan, Huang Maoxian 110514.
Guangxi: Tianlin, Lao Shan, elev. 1350 m, shrub, flowers white, Zhang Zhaoqian 11049 (holotype in SCBI).

This species differs from *C. caudata* by having entire leaves, more lateral veins and sepals longer.

Camellia edentata Chang
1. fruiting branch; 2. flower; 3. capsule.

14. *Camellia salicifolia* Champ. ex Benth. in Hook., Journ. Bot. 3: 309. 1851.
   Thea salicifolia (Champ.) Seem., Trans. Linn. Soc. 22: 349. 1859.
   Thea salicifolia var. *warburgii* Kochs in Engler, Bot. Jahrb. 27: 583. 1900.
   Camelliastrum salicifolium (Champ.) Nakai, Journ. Jap. Bot. 16: 701. 1940.
   Camellia salicifolia var. *longisepala* Keng, Taiwania 1: 233. 1950.
   Branchlets very villous. Leaves thin, lanceolate, pubescent. Sepals linear-lanceolate, 1-1.5 cm long; filaments connate into a tube, pubescent; ovaries and styles pubescent.
   Guangdong: Mei Xian, Zeng Huaide (W. T. Tsang) 21486; Dabu, Zeng Huaide (W. T. Tsang) 21054; Huaiyang, Zeng Huaide (W. T. Tsang) 25636; Conghua, Zeng Huaide (W. T. Tsang) 20525; Renhua, Zeng Huide (W. T. Tsang) 23129; Luoding, Wang Bosun and Xu Linqing 960.
   Hong Kong: Damao Shan, Chen Nianqu 40118.
   Guangxi: Zhaoping, Liang Shengye (S. Y. Liang) 6505249.
   Fujian: Tang Ruiyong 669; southwestern Fujian, J. L. Gressitt 1727.
   Distribution: Guangdong, Guangxi, Fujian, Taiwan.
   There is a great deal of variation in the sepals of this species. The Hong Kong specimens exhibit two types of sepals some 1.5 cm long and some less than 1 cm long although the leaf shapes are all the same. So it is difficult to subdivide these specimens by sepal length. Some specimens of *C. cordifolia* have thin leaves and rounded sepals. Their leaves are similar to the Hong Kong specimens, and they may be hybrids. This type also occurs in Taiwan.

Index

Adinandra 13
Aquifoliaceae 18
Calpandria 27, 136
  lanceolata 136
  quiscosaura 136
Camellia 1, 2, 3, 4, 5, 6, 7, 8, 13, 16, 18, 19, 24, 25, **27**, 56
  acutisepela 163, 166
  acutiserrata 112, **116**
  acutissima 158, 159, **169**, 170, *170*
  albescens 74, **86**
  albogigas 18, 31, **32**, *33*
  albovillosa 74, *85*, **86**
  amplexicaulis 1, **125**
  amplexifolia 1, 117, **120**
  angustifolia 138, **147**, *148*
  anlungensis **67-70**
  assimiloides 1, 200, **204-205**
  assimilis 2, 6, 184, 200, **205**
  aurea 128, **129**
  axillaris 204
  boreali-yunnanica 76, **102-104**
  brachyandra 14, 117, **120-122**, *121*
  brevissima 49, **56-57**
  brevistyla 6, 49, **51**, *53*, 54, 56, 57, 60
  buisanensis 204
  buxifolia 158, 166-168
  callidonta 158, *172*
  campanisepala 160, **189**, *190*, 200
  candida 200, **202-204**
  caudata 2, 6, 14, 172, 199, 200, 201, 202, *203*, **204**, 205
    var. *gracilis* 204
  cavaleriana 94
  chekiangoleosa 6, 7, 25, 76, **104**, *105*, 108
  chrysantha 6, 7, 8, 14, 25, 128, **129**, 130, *131*
    var. *macrophylla* 132
    var. *microcarpa* 129
  chrysanthoides 128, **132**
  chungkingensis 63, **64**, *65*
  compressa 76, **99**, *100*
    var. compressa **99**
    var. variabilis **99**
  confusa 1, 49, **50**
  connata 6, 136, **137**
  corallina 14, 111, **112**
  cordifolia 200, **202**, *203*, 207
  costata 138, **144**
  costei 158, *174*, **175**
  crapnelliana 2, 6, 25, 40, **43**, *46*
  crassicolumna 137, **142**
  crassipes 158, **163**

crassipetala 117, **122**, 123
cratera *183*, 199, **200-201**, 202
crispula 138, **143**
cryptoneura 75, **97**, *98*, 102
cuspidata 6, 14, 157, **162**, 163, 168, 169, 182, 188
  var. chekiangensis 158, **163**
  var. cuspidata 157, **162**
  var. grandiflora 158, **163**
dormoyana 2, 13, 32, **34**
drupifera 3, 38
dubia 160, **187**, *187*
edentata **205**, *206*
edithae 2, 75, **95**, 200
elongata 159, **181**, *182*
euphlebia 1, 128, **132**, *133*
euryoides 4, 5, 159, 168, **178**, 196
  var. *nokoensis* 196
fangchensis 138, **150-151**
flava 1, **128-129**
flavida 128, **129-130**
fleuryi 2, 112, **113**
fluviatilis 49, **51**, *52*
forrestii 6, **166**, *167*, 172
  var. acutisepala 158, 163, **166**
  var. forrestii 158, **166**
fraterna 2, 6, 160, 172, **184-187**, *186*, 189
furfuracea 2, 6, 14, 37, 40, 41, **43-45**, 48
  var. *lutea* 43, 45
gauchowensis 2, 25, **38**
gaudichaudii 40, **45**
gigantocarpa 43
gilbertii 1, 117, **122**
glaberrima 14, *126*, **154-155**
gnaphalocarpa 51
gracilipes 1, 153, **154**
gracilis 204
granthamiana 1, 2, 6, 7, 8, 13, 18, 30, **31-32**, 37
grijsii 2, 6, 7, 25, 48, **50**
gymnogyna 138, 143, 144, *145*, 146
handelii 158, **172**, *173*, 175, 184
havaoi 108
henryana 2, 63, 64, **66**, *68*
henryana 64
heterophylla 92, 94
hiemalis 2, 5, 6, 7, 75, **95**, *103*
hongkongensis 2, 6, 7, 8, 75, 92, **95-97**, *96*
hongkongensis 45
hozanensis 107
ilicifolia 63, **66**

impressinervis 128, **130**, 134, 136
indochinensis 1, 125, **127**, *127*
integerrima 40, **41**, *42*
irrawadiensis 6, 137, 141, 142, **143**
japonica 1, 3, 4, 5, 6, 7, 8, 14, 25, 27, 37, 74, 76, 95, *96*, 104, **107**, 108
  var. *hozanensis* 107
  ssp. japonica
    var. japonica **107**
    var. macrocarpa **108**
  ssp. rusticana **108**
  var. *spontanea* 108
japonica 94, 95
kissii 4, 6, 14, 37, 48, **50**, *52*
  var. kissii 49, **50**
  var. megalantha 49, **51**
  var. *stenophylla* 51
krempfii 1, 18, **32-34**, *33*
kwangtungensis 154, **155**, *156*
kwangsiensis 137, **139**, *140*
kweichowensis 74, **83**, *84*
lanceolata 14, **136-137**
lancicalyx 160, **193-195**
lancilimba 2, 160, **189-192**, *191*, 200
lapidea 74, 77, **80**, *81*
latilimba 43
latipetiolata 2, 40, **43**
lawii 1, 199, **201**
leptophylla 138, **146**
liberistyla 32, **34-36**
liberistyloides 32, **36**
lienshanensis 112, **114**, *115*
litchii 67, **71**
liuii 166
longicalyx 158, 159, **163-166**, *165*
longicarpa 159, **181-182**, *183*
longicaudata 76, *81*, **108-109**
longicuspis 158, 159, **163**, *164*
longipedicellata 14, 25, **125**, *126*
longissima 14, **153-154**
lucidissima 76, **104**, *106*
lungshenensis 75, 76, *91*, **92**
lutchuensis 6, 159, **178**
luteoflora 8, 14, **73-74**
lutescens 2, 49, **51**
macrosepala 157, *161*, **162**
magnocarpa 76, **107**
mairei 77, **80**, *82*
  var. alba 74, **80**
  var. mairei 74, **80**
  var. *lapidea* 80
maliflora 1, 4, 5, 6, 49, **54**

209

mastersia 50
melliana 200, **202**
membranacea 160, **188**
microphylla 2, 49, 51, *59*, **60**
minahassae 136
minutiflora 158, **168**
miyagii 2, 6, 49, **56**
multiperulata 75, **90-92**, *91*
muricatula **117-118**
nematodea 1, 117, **122**
nervosa 2, 117, **122**
nitidissima 2, 111, 112, **113**
nokoensis 6, 161, **196**
oblata 40, *44*, **45**
obovatifolia 67, **70**
obscurinervis 200
obtusifolia 49, **54**, *55*, 56, 57
octopetala 6, 43
oleifera 1, 3, 4, 5, 6, 7, 14, 25, 37, 38, **40**, 50
  var. *confusa* 1, 50
oleosa 40
omeiensis 74, **77**, *78*
oviformis 76, **97-99**
pachyandra 2, 117, **118**, *119*, 122
parafurfuracea 40, *47*, **48**
parvicaudata 160, *194*, **195**
parvicuspidata 158, **168-169**
parviflora 1, 6, 117, 118, 120, **123**, *124*
parvifolia 184
parvilapidea 159, 184, *185*
parvilimba 180, *180*
  var. brevipes 159, **181**
  var. parvilimba 159, **180-181**
parvimuricata 67, **71-72**, *72*
parviovata 160, **192-193**
parvisepala 138, **151-153**, *152*
paucipunctata 2, 37, 112, **113**
pentamera 112, **114**, 116
pentastyla 137, 141, **142**
percuspidata 160, 163, **188**
petelotii 1, **125**
phaeoclada 49, **60**, *61*
phellocapsa 75, **88-90**, *89*, 99
pilosperma 111, **113**, *119*
pingguoensis 128, **134**, *135*
piquitiana 2, 14, **36**
pitardii 1, 6, 7, **94**, 102
  var. alba 74, **94-95**
  f. *cavaleriana* 94
  var. *lucidissima* 102
  var. pitardii 75, **94**
  var. yunnanica 7, 76, **94**, 102
pleurocarpa 1, 18, 31, **32**
polyodonta 74, 77, *79*, 80, 90, 92, **97**
polypetala 40, **41-43**, *44*
ptilophylla 24, 138, 150, **151**
pubicosta 138, 139, **147**
pubipetala 128, **134-136**
pubisepala 43, 45
punctata 1, 199, **201**
puniceiflora 49, 57, *59*
quinquelocularis 137, 138, **139**

reticulata 1, 3, 4, 6, 7, 25, 75, 76, 83, 86, **92-94**, *93*
  var. *albo-rosea* 6
  var. *campanulata* 6
  var. *rosea* 6, 95
  var. *wabiske* 6
rhytidocarpa 67, **70**, 71
rosaeflora 5, 6, 160, **188-189**
rosthorniana 159, 166, **178**, *179*
rusticana 6, 108
salicifolia 2, 6, 200, **207**
  var. *longisepala* 207
saluenensis 6, 7, 76, 80, 94, **102**, *103*, 202
  f. *minor* 102
sasanqua 1, 3, 4, 5, 6, 7, **38**, 56
sasanqua 54
scariosisepala 112, **116**
scottiana 25, 149
semiserrata 2, 25, 88, 90, 107
  var. albiflora 74, **90**
  var. *magnocarpa* 107
  var. semiserrata 75, **90**
setiperulata 76, *101*, **102**
shensiensis 7, 49, **56**, *58*
sinensis 1, 3, 6, 7, 14, 111, 137, 144, 146, **147**, 151
  var. assamica 24, 25, 138, **149**
  var. pubilimba 138, **150**
  var. sinensis 24, 25, 138, **147-149**
  f. *macrophylla* 147
  f. *parvifolia* 147
  var. waldenae 138, **150**
speciosa 102
stenophylla 51
stuartiana 159, 172, **184**
subacutissima 158, 159, **169-170**, 172, *171*
subglabra 160, 193, **196**
subintegra 76, *106*, 108
symplocifolia 50
synaptica 178
szechuanensis 14, 63, 64, *69*
szemaoensis 117, **118**
tachangensis 137, **141**
taheishanensis 118
taliensis 6, 137, 141, 142, **143**
tenii 2, 37, 49, **57**, *59*
tenuiflora 51
tetracocca 141
thea 147
theiformis 178
tonkinensis 1, 111, **112**
transarisanensis 6, 160, **184**
transnokoensis 6, 159, **178**
triantha 158, **172-175**
trichandra 161, *198*, **199**
trichocarpa 63, **64-66**, 200
trichoclada 159, *176*, **179-180**
trichosperma 75, 88
trigonocarpa *197*, 199, **201-202**
tsaii 6, **175**, *176*
  var. synaptica 159, *177*, **178**, 182
  var. tsaii 159, **175**

tsingpienensis **192**, 193
  var. pubisepala 160, **192**
  var. tsingpienensis 160, **192**
tsofuii 161, **196**, *197*, 199
tuberculata 1, 14, *67*, *69*, 70, 71
tunganica 74, **86-88**, *87*
tunghinensis 128, 129, **132-134**, *133*
uraku 2, 5, 6, 7, 75, **95**
vernalis 5, 6
vietnamensis 6, 25, **38**, *39*
villicarpa 1, *197*, 199, **200**
villosa 74, 77, **80-83**
viridicalyx 160, **193**, *194*
wabiske 5, 6
waldenae 150
wardii 2, **112-113**
weiningensis 49, **60-63**
wenshanensis 200, **202**
yangkiangensis 117, **123**
yuhsienensis 6, 7, 50
yungkiangensis 138, *145*, **146**
yunnanensis 2, 6, 7, 16, 18, 32, **34**, *35*, 37
X maliflora 6
X vernalis 6
xylocarpa 75, 76, **95**
xanthochroma 117, **118-120**

Camelliastrum 27
  assimile 205
  buisanense 204
  caudatum 202
  gracile 204
  mairei 80
  salicifolium 207
Demitus 27
  reticulata 92
Dilliales 16
Drupifera 27
Ericaceae 18
Eurya 13
Glyptocarpa 18, 27
  camellioides 34
Fagaceae 18
Gordonia
  yunnanensis 143
Kemelia 27
Kailosocarpus 27
  camellioides 34
Lauraceae 18
Moraceae 18
Myrtaceae 18
Parapiquetia 27
Piquetia 27
  piquetiana 36
Polyspora
  yunnanensis 143
Pyrenaria
  camelliaeflora 136
  camellioides 34
Rosaceae 18
Salceda
  montana 136
Sasanqua 5, 27
  malliflora 54

INDEX

odorata 6, 38
vulgaris 6, 38
Section
   Archecamellia 1, 2, 13, 15, 16, fig. 1, 18, tab. 1, 27, **31**, 32
   Brachyandra 1, 2, 14, 15, 16, fig. 1, 19, tab. 1, 28, **116-117**, 122
   Calpandria 14, 15, 16, fig. 1, 19, tab. 1, 28, **136**
   Camellia 2, 5, 7, 14, 15, fig. 1, 18, 19, tab. 1, 28, 34, **74,** 83, 92, 95
   Camelliopsis 1, 2, 199
   Chrysantha 1, 7, 8, 14, 15, 16, fig. 1, 18, 19, tab. 1, 28, 73, 111, **128**
   Corallina 1, 2, 14, 15, 16, fig. 1, 19, tab. 1, 28, **111,** 114, 117
   Eriandria 1, 2, 14, 15, fig. 1, 19, tab. 1, 28, **199**
   Furfuracea 2, 7, 14, 15, fig. 1, 19, tab. 1, 25, 28, 37, **40,** 41, 43
   Glaberrima 14, 15, 16, fig. 1, tab. 1, 28, **154**
   Heterogena 1, 2, 37
   Longipedicellata 1, 14, 15, 16, fig. 1, tab. 1, 28, **125**
   Longissima 1, 14, 15, 16, fig. 1, tab. 1, 28, **153**
   Luteoflora 14, 15, fig. 1, tab. 1, 28, **73**
   Oleifera 1, 7, 14, 15, fig. 1, 19, tab. 1, 25, 28, **37-38,** 40, 48
   Paracamellia 1, 2, 6, 7, 14, 15, fig. 1, 19, tab. 1, 25, 28, 37, **48, 54,** 73
   Piquetia 14, 15, fig. 1, tab. 1, 28, **36**
   Pseudocamellia 1, 2, 14, 15, fig. 1, 19, tab. 1, 28, **63**
   Stereocarpus 1, 2, 13, 15, 16, fig. 1, 18, tab. 1, 28, **32**
   Thea 1, 14, 15, 16, fig. 1, 19, tab. 1, 24, 28, **137,** 154
   Theopsis 1, 14, 15, fig. 1, 19, tab. 1, 28, 54, **157,** 172, 188, 189
   Tuberculata 1, 14, 15, fig. 1, 19, tab, 1, 28, **67**
Series
   Chrysanthae 128, **129**
   Cuspidatae 157, **162**
   Flavae **128**
   Gymnandrae 158, **172**
   Gymnocarpae 63, **66**
   Gymnogynae 138, **143**
   Quinquellocularis 137, **138**
   Pentastylae 137, **141**
   Reticulatae 74, **83**
   Sinenses 138, **146**
   Trichandrae 160, **192**
   Trichocarpae **63**
   Villosae 74, **77**
Stereocarpus 27
   dormoyanus 34
Subgenus
   Camellia 7, 8, 14, 15, 16, fig. 1, 18, 19, tab. 1, 28, 34, **37,** 95
   Metacamellia 7, 8, 14, 15, 16, fig. 1, 18, 19, tab. 1, 28, **157**
   Protocamellia 7, 13, 14, 15, 16, fig. 1, 18, tab. 1, 27, **31,** 37, 83
   Thea 7, 8, 14, 15, 16, fig. 1, 18, 19, tab. 1, 28, 37, 73, **111**
Subsection
   Lucidissima 19, 76, **104**
   Reticulata 74, **76,** 90
Symplocaceae 18
Ternstroemia 13
Ternstroemioxylon 19
Thea 6, 27, 111, 137
   amplexicaulis 125
   assamica 149
   assimilis 205
   bachmaensis 50
   bolovensis 43
   bohea 147
   brachystemon 50
   brevistyla 51
   buisanensis 204
   camellia 107
      var. *lucidissima* 102
   cantonensis 147
   caudata 204
      var. *faberi* 181
   cavaleriana 94
   chinensis 147
   cochinchinensis 147
   confusa 50
   connata 137
   corallina 112
   cordifolia 202
   costei 175
   cuspidata 162
   dormoyana 34
   edithae 95
   elongata 181
   euryoides 178
   flava 128
   fleuryi 113
   fluviatilis 51
   forrestii 166
   fraterna 172
   furfuracea 43
   fusiger 175
   gaudichaudii 45
   gilbertii 122
   gnaphalocarpa 51
   grijsii 50, 94
   henryana 66
   hongkongensis 95
   hozanensis 107
   indochinensis 127
   iniquicarpa 50
   japonica 107
   krempfii 32
   lutchuensis 178
   lutescens 51
   mairei 80
   maliflora 54
   melliana 202
   microphylla 60
   miyagii 56
   nematodea 122
   nervosa 122
   nokoensis 196
   oleosa 147
   parvifolia 184
   paucipunctata 113
   petelotii 125
   piquitiana 36
   pitardii 80, 94, 102
      var. *lucidissima* 102
   pleurocarpa 32
   podogyna 40
   polygama 166
   punctata 201
   reticulata 92
      var. *rosea* 95
   rosaeflora 188
      var. *glabra* 162
      var. *pilosa* 184
   rosthorniana 178
   salicifolia 207
      var. *warburgii* 207
   sasanqua 38
      var. *kissii* 50
   sinensis 5, 147
   sp. 67
   taliensis 143
   tenuiflora 51
   theiformis 178
   tonkinensis 112
   transarisanensis 184
   transnokoensis 178
   trichoclada 179
   tsaii 175
   viridis 147
      var. *assamica* 149
   yunnanensis 34
Theaceae 13, 16, 19
Theaphylla 27
   euryoides 178
Theopsis 27, 157
   amplexifolia 120
   caudata 204
   chrysantha 129
   elongata 181
   euonymifolia 57
   forrestii 166
   fraterna 184
   furfuracea 43
   longipedicellata 125
   lungyaiensis 51
   lutchuensis 178
   maliflora 54
   microphylla 60
   nokoensis 196
   parvilimba 180
   polygama 166
   transarisanensis 184
   transnokoensis 178
   trichoclada 179
Tsia 27
Tsubaki 27
Yunnanea 27
   xylocarpa 95